The Open University

Science Foundation Course Unit 10

COVALENT COMPOUNDS

Prepared by the Science Foundation Course Team

THE OPEN UNIVERSITY PRESS

A NOTE ABOUT AUTHORSHIP OF THIS TEXT

This text is one of a series that, together, constitutes *a component part* of the Science Foundation Course. The other components are a series of television and radio programmes, home experiments and a summer school.

The course has been produced by a team, which accepts responsibility for the course as a whole and for each of its components.

THE SCIENCE FOUNDATION COURSE TEAM

M. J. Pentz   (Chairman and General Editor)

F. Aprahamian	(Editor)	A. R. Kaye	(Educational Technology)
A. Clow	(BBC)	J. McCloy	(BBC)
P. A. Crompton	(BBC)	J. E. Pearson	(Editor)
G. F. Elliott	(Physics)	S. P. R. Rose	(Biology)
G. C. Fletcher	(Physics)	R. A. Ross	(Chemistry)
I. G. Gass	(Earth Sciences)	P. J. Smith	(Earth Sciences)
L. J. Haynes	(Chemistry)	F. R. Stannard	(Physics)
R. R. Hill	(Chemistry)	J. Stevenson	(BBC)
R. M. Holmes	(Biology)	N. A. Taylor	(BBC)
S. W. Hurry	(Biology)	M. E. Varley	(Biology)
D. A. Johnson	(Chemistry)	A. J. Walton	(Physics)
A. B. Jolly	(BBC)	B. G. Whatley	(BBC)
R. Jones	(BBC)	R. C. L. Wilson	(Earth Sciences)

The following people acted as consultants for certain components of the course:

J. D. Beckman	B. S. Cox
H. G. Davies	R. J. Knight
D. J. Miller	M. W. Neil
J. R. Ravetz	H. Rose

The Open University Press
Walton Hall, Bletchley, Bucks

First published 1971 Reprinted 1972
Copyright © 1971 The Open University

Designed by The Media Development Group of the Open University

Printed in Great Britain by
EYRE AND SPOTTISWOODE LIMITED
AT GROSVENOR PRESS PORTSMOUTH

SBN 335 02004 6

Open University courses provide a method of study for independent learners through an integrated teaching system, including textual material, radio and television programmes and short residential courses. This text is one of a series that make up the correspondence element of the Science Foundation Course.

The Open University's courses represent a new system of university level education. Much of the teaching material is still in a developmental stage. Courses and course materials are, therefore, kept continually under revision. It is intended to issue regular up-dating notes as and when the need arises, and new editions will be brought out when necessary.

For general availability of supporting material referred to in this book, please write to the Director of Marketing, The Open University, Walton Hall, Bletchley, Buckinghamshire.

Further information on Open University courses may be obtained from The Admissions Office, The Open University, P.O. Box 48, Bletchley, Buckinghamshire.

1.2

Contents

	List of Scientific Terms, Concepts and Principles	5
	Conceptual Diagram	6
	Objectives	7
10.1	**Introduction**	9
10.1.1	Summary of section 10.1	10
10.2	**Covalency in Carbon Compounds**	12
10.2.1	Summary of section 10.2	16
10.3	**The Uniqueness of Carbon**	17
10.3.1	Summary of section 10.3	22
10.4	**The Shapes of Organic Molecules**	23
10.4.1	Effects of electron repulsion	23
10.4.2	Molecules which contain single bonds only	26
10.4.3	Molecules which contain multiple bonds	27
10.4.4	Naturally occurring molecules	31
10.4.5	Chirality	35
10.4.6	Summary of section 10.4	38
10.5	**The Influence of Structure on Physical and Chemical Properties**	39
10.5.1	Introduction	39
10.5.2	Melting point and boiling point sequences	39
10.5.3	Solubility	41
10.5.4	Colour	42
10.5.5	Chemical reactivity	44
10.5.6	Summary of section 10.5	46
10.6	**The Influence of Structure on Properties — Examples from Chemical Technology**	47
10.6.1	Motives and methods of chemical technology	47
10.6.2	Dyes	48
10.6.3	Polymer chemistry	49
10.6.4	Pharmaceutical chemistry	50
10.6.5	Agricultural chemistry	51
10.6.6	Detergents	52
10.6.7	New problems	53
10.6.8	Summary of section 10.6	54
10.7	**Summary**	55
	Book List	57
	Appendix 1 (White) Ball and spring molecular models	58
	Appendix 2 (White) Plane-polarized light	60
	Appendix 3 (Black) The orbital approach to the geometry of organic molecules	61
	Appendix 4 (Black) Distortion factors of electron repulsion theory	62
	Appendix 5 (Black) The structure of benzene	63
	Appendix 6 (Black) Electronic structure and chemical reactivity	66
	Appendix 7 (Black) Nomenclature in organic chemistry	71
	Self-Assessment Questions	75
	Self-Assessment Answers and Comments	86
	Answers to In-text Questions	93

Table A

A List of Scientific Terms,* Concepts and Principles Used in Unit 10

Taken as prerequisites			Introduced in this Unit			
1	**2**		**3**		**4**	
Assumed from general knowledge	Introduced in a previous unit	Unit No.	Developed in this unit	Page No.	Developed in a later unit	Unit No.
natural gas	Periodic		polarization	9	carbohydrate	13
flint, quartz	Table,	7, 8	valence shell	12	protein	13
evolution	covalent and		structural isomers	14	polysaccharides	13
melting point	ionic com-		double, triple and multiple		enzymes	15
boiling point	pounds	8	bonds	15	polymers	13
dye	electronic		valency number	18		
anaesthetic	structure	7	non-bonding electronic pairs	18		
antibiotic	carbon		organic chemistry	21		
pesticides	tetrachloride	8	bond angle	24		
DDT	noble gas	8	stereochemistry	24		
nerve gas	halogen	8	conformation	26		
toxicity	Pauli exclusion		saturated and unsaturated			
detergent	principle	7	molecules	26, 27		
soap	quantum		stereoisomerism	28		
	numbers	6	geometric isomerism	28		
	electron spin	7	polarizability	29		
	solvent	9	monosaccharides	31		
	wavelength	2	amino acid	34		
	aqueous	9	chirality	35		
	ions	9	plane of symmetry	36		
			achiral	36		
			plane-polarized light	37		
			optical isomers	37		
			hydrogen bonding	41		
			conjugated system	42		

* Excluding the names of chemical compounds.

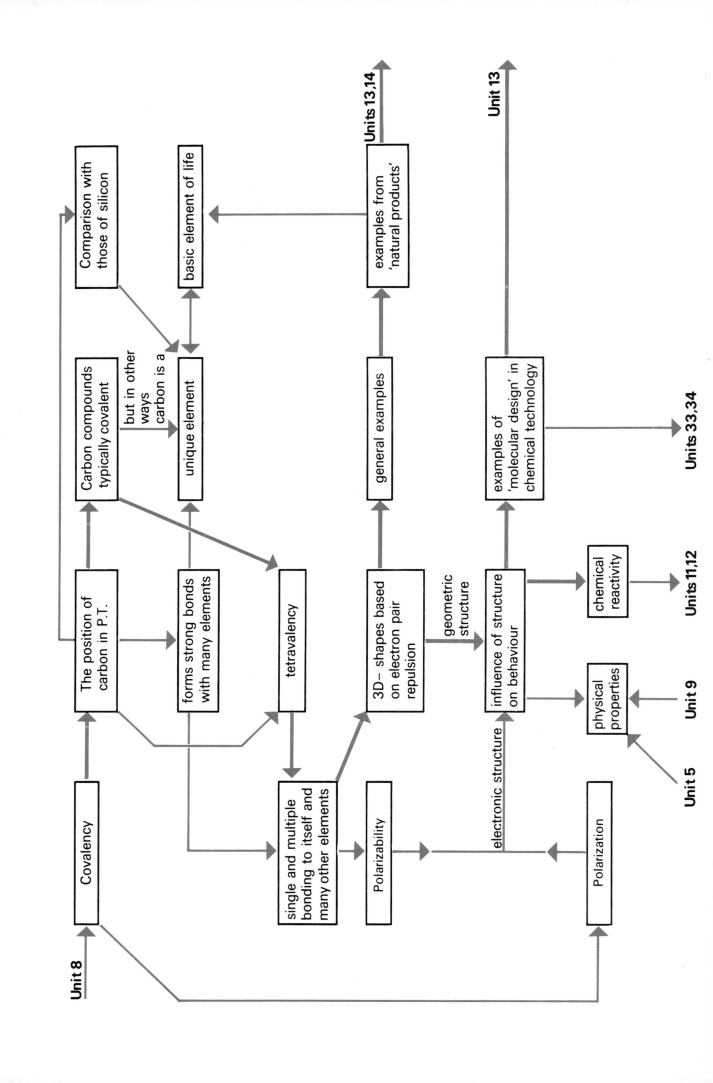

Objectives

The self-assessment questions (*SAQ*) in which these objectives are tested are given in brackets after each objective. An objective may also be tested by a question appearing in the text (*TQ*).

1. Define in your own words, recognize valid definitions of, or use in a correct context the terms or expressions in column 3 of Table A.
(*SAQs* 1, 2, 8, 15, 17, 22)

2. Write down all the two-dimensional structural formulae possible for a given molecular formula C_xH_y where x and y are integers up to and including five and twelve respectively.
(*TQ*)

3. Write down and identify correct representations of methane, ethane, propane, ethylene and acetylene.
(*SAQ* 3)

4. Recognize valid two-dimensional representations of structures of organic molecules which contain, in addition to carbon and hydrogen, one or more of the elements nitrogen, oxygen, fluorine, chlorine, bromine and iodine.
(*SAQ* 4)

5. Given the molecular formula of a compound of the type described under 4 above, write down possible two-dimensional structural formulae.
(*TQ*)

6. Describe reasons (in less than 300 words) why the compounds of carbon and silicon play different roles in nature.
(*SAQs* 5, 7)

7. Identify structural isomeric relationships among a given series of structural formulae.
(*TQ*)

8. Give reasons in order of importance why carbon forms more covalent compounds than all other elements combined.
(*SAQ* 6)

9. Explain (in less than 50 words) the origin of the term 'organic chemistry'.
(*SAQ* 8)

10. Use electron repulsion theory to deduce the approximate shape of a simple covalent molecule.
(*SAQ* 11)

11. State the influence of bond angles in a molecular fragment if;
    1 bonding electron pairs are replaced by non-bonding electron pairs;
    2 one of the bonds is a double bond;
    3 there is a large difference in electronegativity between the bonded atoms.
(*SAQs* 9, 12)

12. Recognize valid two-dimensional representations of the three-dimensional structures of organic compounds with less than ten carbon atoms.
(*SAQ* 13)

13. Identify features in a molecule which reduce its structural flexibility.
(*TQ, SAQ* 17)

14.   Draw correct three-dimensional representations of ethane, ethylene, acetylene, glucose and a fragment of the primary structure of a protein. (*TQ, SAQ* 16)

15.   Recognize geometrical and optical isomerism within a series of structural isomers.
(*SAQ* 14)

16.   Classify any object or simple molecule as being chiral or achiral.
(*TQ, SAQ* 15)

17.   Classify each of the organic compounds given in Objective 18 as a carbohydrate, amino acid, or neither of these.
(*SAQ* 15)

18.   Demonstrate knowledge of the structures of diethyl ether, glycine, alanine, tetrahydrofuran, glucose and mannose by identifying features which are present or absent in their structures.
(*SAQ* 15)

19.   Relate gross trends in melting points, boiling points, solubilities (in a given solvent) and colour within a given series of compounds to the concepts of polarization, polarizability, conjugation, hydrogen-bonding and molecular shape.
(*TQ, SAQs* 17, 18, 19, Home Experiment)

20.   Recognize simple structural features which may give rise to chemical reactivity.
(*SAQ* 21)

21.   Demonstrate an understanding of the reason why an enzyme normally accepts only one of a mirror-image pair of molecules for catalytic chemical change, by drawing relevant diagrammatic models and by recognizing valid analogies.
(*SAQ* 20)

22.   Recognize some typical molecular structures associated with each of the following fields of chemical technology:
>   petroleum and petrochemicals
>   dyes
>   pharmaceuticals
>   pesticides
>   polymers
>   detergents.
(*SAQ* 23)

23.   Give at least two examples of molecular design in chemical technology.
(*SAQ* 24)

24.   Identify concepts in the Unit that are relevant to:
>   the role of carbon compounds in nature;
>   molecular shape;
>   chemical reactivity;
>   physical properties;
>   molecular design in chemical technology and its impact on
>   society;
and assess the extent of this relevance.

10.1 Introduction

Units 6, 7 and 8 have progressively developed a picture of the structure of matter at the atomic level which rationalizes a classification of the elements according to their properties in a form known as the Periodic Table. Furthermore, in Unit 8 you saw that this classification and its supporting theoretical rationale is immensely useful in predicting the types of chemical compounds a particular element would form. It is apparent, for instance, that atoms of elements positioned near the centre of the Table (Group IV for example) are expected to form bonds with other atoms by electron sharing. Those elements situated at either edge, on the other hand, should bond to other atoms by electron donation (Group I) or by accepting an electron (Group VII). This gives rise to the two important classes of materials: covalent and ionic compounds respectively. You recall the examples given in Unit 8.

$$Cl-\underset{\underset{Cl}{|}}{\overset{\overset{Cl}{|}}{C}}-Cl$$ carbon tetrachloride (covalent) $Na^{+}Cl^{-}$ sodium chloride (ionic)

Unit 9 explored the properties of ionic compounds and the present Unit sets out to examine those of covalent compounds. Before we begin to focus attention on the latter, however, it is important that you should bear in mind that the dividing line between these two types is by no means sharp. A large number of compounds are essentially covalent in that electrons are being shared between atoms, but at the same time they are partially ionic because one of a pair of atoms is claiming a greater share of the binding electrons than the other. This situation produces *polarization*, that is, an unequal distribution of charge between the two atoms.

polarization

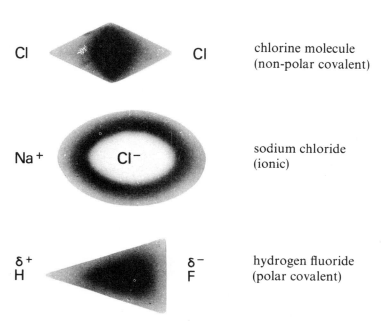

Cl ⬥ Cl   chlorine molecule (non-polar covalent)

Na⁺ (Cl⁻)   sodium chloride (ionic)

δ^{+} H ◀ δ^{-} F   hydrogen fluoride (polar covalent)

Figure 1   Ionic and covalent electron distributions (δ is a symbol we use to denote charge intermediate between zero and a full unit; the shading denotes the approximate distribution of the bonding pair of electrons).

You will see later that polarity within a covalent molecule exerts a powerful influence on its behaviour.

Our consideration of covalent compounds in this Unit will be restricted to elements of Group IV of the Periodic Table (Fig. 2) and we shall focus attention largely on the first member of Group IV, carbon. We have chosen to concentrate on carbon for three reasons. First, the number of known covalent compounds involving carbon greatly exceeds the number of all other covalent compounds put together. Secondly, the physical and chemical properties of covalent compounds are typified by those involving carbon. Thirdly, carbon occupies a uniquely important role in nature, being the basic element of all living systems.

We shall start by exploring the types of compounds carbon forms with other elements, and a large part of the Unit will be concerned with their shapes.

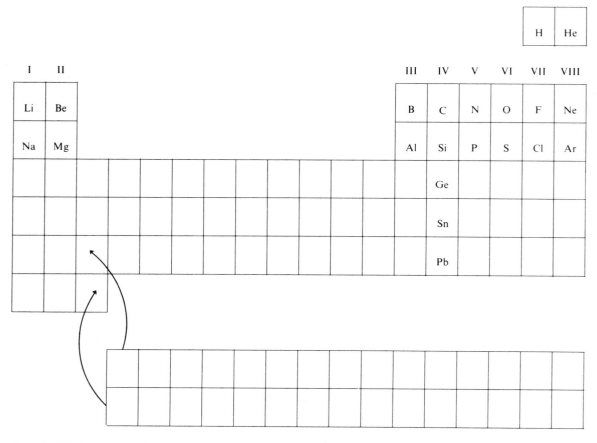

Figure 2   The location of carbon and Group IV in the Periodic Table.

Once we have established their geometric structures we shall return to electron distribution and then consider how both these aspects of structure control chemical and physical properties. The Unit ends with a brief description of the manner in which increasing knowledge of this structure/property relationship at the molecular level is rapidly changing our way of life.

10.1.1   Summary of section 10.1

Bonds in molecules range from the purely covalent, in which electrons are equally shared between two atoms, through polar covalent-bonds, in which they are shared but not equally, to purely ionic where one of the two atoms donates an electron to the other completely.

Study comment

The explanation of the concepts and principles dealt with in Unit 10 relies heavily on the use of examples. This has made it physically rather long and it is important that you should realize at the outset that you do not have to remember all the detailed material it contains, but should concentrate on mastering the principles those details illustrate. You should therefore pay particular attention to the objectives, the study comments found at the beginning of some sections, and the summaries at the end of every section.

We have followed the usual practice of placing the answers to the seeded questions in the right-hand margin immediately following the question. In a few instances, particularly where structures are involved, you may be unable to avoid seeing some of the answer out of the corner of your eye whilst reading the question. In order to avoid this possible frustration, we suggest you have a piece of paper or card handy to cover up such answers.

At the end of each section you are asked to summarize what you consider to be the most important points, before reading the summary we have made. You might bear this in mind as you are reading the text, perhaps jotting down the main points as you go along. You will also find at the end of each section a reference to relevant self-assessment questions, answers and comment. This information is for those of you who like to test your understanding of the material section by section. Please do not think we expect you to have attempted *all* of these questions by the end of the week !

The introduction of new chemical names has been avoided wherever possible. The nomenclature of carbon compounds is the subject of a black-page Appendix (Appendix 7, p. 71).

10.2 Covalency in Carbon Compounds

Study comment

> All the structural formulae drawn in this section are used to illustrate the kinds of structures possible for compounds of carbon and hydrogen. You need only remember those mentioned in Objective 3. After reading it, however, you should be able to achieve Objective 2 and know the meanings of the terms valence, electrons, hydrocarbons, and structural isomers.

If you refer back to Appendix 1 of Unit 7 you will find that the electronic structure of carbon is $1s^2, [2s^2 2p^2]$ where the valence shell (highest occupied shell) is denoted by the square brackets. It needs to acquire four more electrons to achieve the stable noble gas configuration of neon, $1s^2[2s^2 2p^6]$. It can do this by pairing each of its four valence electrons with one from another atom so that it is surrounded by four pairs of electrons in its valence shell. You have seen this structure in the molecule carbon tetrachloride, CCl_4. The same pattern is shown by methane, the simplest stable compound that carbon forms with hydrogen.

$$.\overset{.}{\underset{.}{C}}. \;+\; 4H\text{×} \;=\; H\overset{\overset{H}{\underset{}{\text{×}}}}{\underset{\underset{H}{\text{×·}}}{\text{×}\underset{.}{C}\text{×}}}H \quad \text{or} \quad H-\overset{\overset{\textstyle H}{|}}{\underset{\underset{\textstyle H}{|}}{C}}-H \qquad \text{methane}$$

Here, we have shown how the electronic structure of methane is built up from one carbon atom and four hydrogen atoms and, for the sake of illustration, have distinguished between the valence structure electrons of carbon (·) and hydrogen (x), although, as you will see later in Unit 29, such a distinction is unrealistic once the compound is formed.

The hydrogen atoms achieve the stable electronic structure of helium. The C—H bonds so formed are usually very difficult to break; that is, they are very strong. Carbon can also form strong bonds with other carbon atoms. Thus, stable compounds result when carbon and hydrogen atoms are mutually bound in one molecule in such a way that each atom attains effectively a noble gas electronic structure.

2 carbon atoms

$$H\overset{\overset{H\;\;H}{\text{×· ×·}}}{\underset{\underset{H\;\;H}{\text{×· ·×}}}{\text{×}C\text{:}C\text{×}}}H \;\equiv\; H-\overset{\overset{\textstyle H}{|}}{\underset{\underset{\textstyle H}{|}}{C}}-\overset{\overset{\textstyle H}{|}}{\underset{\underset{\textstyle H}{|}}{C}}-H \qquad \text{ethane}$$

3 carbon atoms

$$H\text{×}\overset{\overset{H\;H\;H}{\text{·× ×· ·×}}}{\underset{\underset{H\;H\;H}{\text{×· ·× ×·}}}{C\text{:}C\text{:}C}}\text{×}H \;\equiv\; H-\overset{\overset{\textstyle H}{|}}{\underset{\underset{\textstyle H}{|}}{C}}-\overset{\overset{\textstyle H}{|}}{\underset{\underset{\textstyle H}{|}}{C}}-\overset{\overset{\textstyle H}{|}}{\underset{\underset{\textstyle H}{|}}{C}}-H \qquad \text{propane}$$

4 carbon atoms

$$H-\overset{\overset{\textstyle H}{|}}{\underset{\underset{\textstyle H}{|}}{C}}-\overset{\overset{\textstyle H}{|}}{\underset{\underset{\textstyle H}{|}}{C}}-\overset{\overset{\textstyle H}{|}}{\underset{\underset{\textstyle H}{|}}{C}}-\overset{\overset{\textstyle H}{|}}{\underset{\underset{\textstyle H}{|}}{C}}-H \qquad \text{and so on} \ldots$$

```
    H   H   H   H   H   H   H   H   H   H   H   H   H
    |   |   |   |   |   |   |   |   |   |   |   |   |
H — C — C — C — C — C — C — C — C — C — C — C — C — C — H
    |   |   |   |   |   |   |   |   |   |   |   |   |
    H   H   H   H   H   H   H   H   H   H   H   H   H
```

The structures of these compounds of hydrogen and carbon (hydro-carbons) are not restricted to linear chains. Branched chains and rings are also possible.

```
              H
              |
    H   H — C — H   H
    |       |       |
H — C ——— C ——— C — H
    |       |       |
    H       H       H
```

```
    H   H   H   H       H       H H — C — H  H H — C — H  H
    |   |   |   |       |       |   |        |   |        |
H — C — C — C — C ——— C ——— C ——— C ——— C ——— C ——— C — H
    |   |   |   | H — C — H    |       |   H H — C — H  H
    H   H   H   H     |        H       H        |
                H — C — H                       H
                    |
                    H
```

In fact, a variety of hypothetical structures of this kind can be built up with any given number of carbon atoms. The carbon atoms can be linked in any direction by two-electron covalent bonds, and hydrogen atoms then added until all the carbon atoms have four bonds. Let us do the exercise with six carbon atoms. Possible structures include:

```
C — C — C              C            C — C — C — C — C — C
|       |              |
C — C — C          C — C — C — C          III
   I                   |
                       C  II
```

```
      C                        C
    C⟋ ⟍C                       |
    |     C — C            C — C — C — C
    C⟍  ⟋                      |
      C   IV                   C  V
```

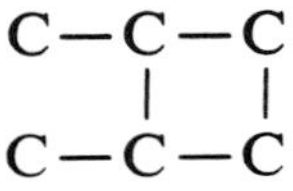

```
C—C—C              C              C—C—C—C—C—C
|   |              |
C—C—C          C—C—C—C                III
   I               |
                   C   II
```

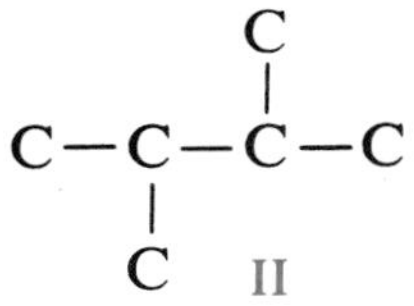

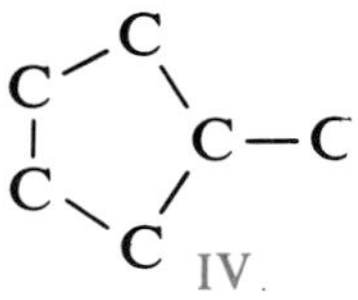

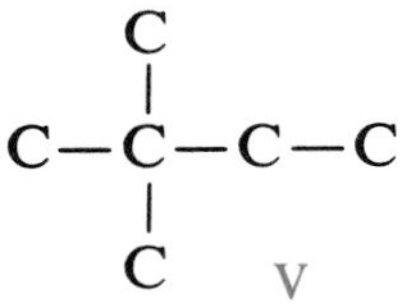

```
    C
  C⁄               C
C               C—C—C—C
|    C—C          |
C   ⁄             C   V
 C
   C   IV
```

(carbon networks I–V shown above)

Write down the complete hydrocarbon structures associated with each of the carbon networks I–V.

See Answer 1, p. 93.

All these structures have six carbon atoms. Count up the numbers of hydrogen atoms in each case and then, on the basis of the result, divide the five molecules into two classes.

You will notice that structures II, III and V have the same molecular formula C_6H_{14}, as do I and IV, C_6H_{12}.

Compounds which possess the same molecular formula but have different structures are called *structural isomers*. The representations of structures which we have been drawing are confined to two dimensions. As you will see shortly, however, the molecules they represent are not flat and, furthermore, the atoms and groups of atoms they contain can rotate about the bonds.

structural isomers

Do not worry if this is not very clear to you at the moment. It means that, for example, the structures VI and VII are not isomers but represent identical molecules. The same applies to VIII and IX.

VI

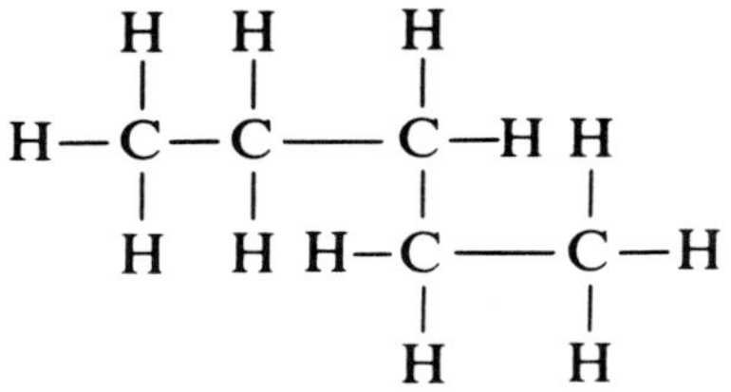

```
    H   H   H
    |   |   |
H—C—C——C—H H
    |   |   |   |
    H   H H—C——C—H
            |   |
            H   H
```

VII

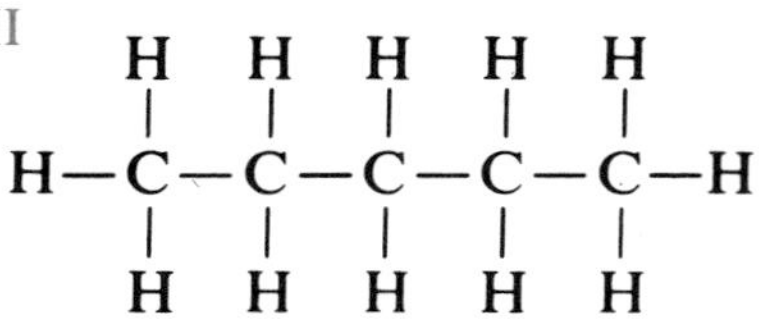

```
    H   H   H   H   H
    |   |   |   |   |
H—C—C—C—C—C—H
    |   |   |   |   |
    H   H   H   H   H
```

VIII

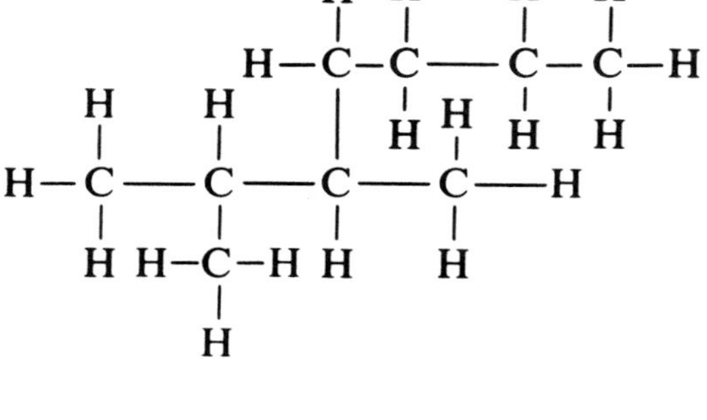

```
            H  H    H  H
            |  |    |  |
        H—C—C——C—C—H
    H   H   |  H H  H  H
    |   |   |
H—C——C——C——C——H
    |   |   |   |
    H H—C—H H   H
        |
        H
```

IX

```
                              H
                              |
    H  H  H  H H—C—H H      H
    |  |  |  |  |     |      |
H—C—C—C—C——C——C——C—H
    |  |  |  |  |     |      |
    H  H  H  H  H H—C—H H
                     |
                     H
```

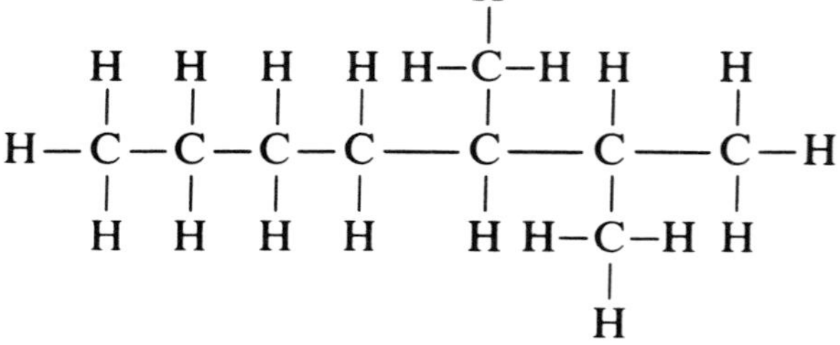

This becomes easier to visualize if we use a simpler notation in which the number of lines in a structural formula are restricted to bonds between carbon atoms only.

14

$$CH_3-CH_2-CH_2 \equiv CH_3-CH_2-CH_2-CH_2-CH_3 \qquad VII$$
$$\underset{VI}{} \quad \underset{\big|}{} CH_2-CH_3$$

VI

$$\underset{\underset{VIII \quad CH_3}{\big|}}{CH_3-CH-CH-CH_3} \quad \underset{\underset{CH_2-CH_2-CH_2-CH_3}{\big|}}{} \equiv CH_3-CH_2-CH_2-CH_2-\underset{\underset{CH_3}{\big|}}{CH}-\overset{\overset{CH_3}{\big|}}{CH}-CH_3$$

IX

Use this notation to write the structural formulae of all the isomers of C_5H_{12}.

See Answer 2, p. 93.

All the structures we have considered so far involve bonds between atoms formed by the sharing of two electrons. Consider again for a moment the formation of a molecule involving two carbon atoms.

$$\cdot \overset{\cdot}{\underset{\cdot}{C}} \cdot \ + \ \underset{\times}{\overset{\times}{_\times}C}{^\times} \ \longrightarrow \ \cdot\overset{\cdot}{\underset{\cdot}{C}} \underset{\times}{\overset{\times}{_\times}}C^\times \ \equiv \ \cdot\overset{\cdot}{\underset{\cdot}{C}}-\underset{\times}{\overset{\times}{C}}\cdot$$

By pairing each of the remaining unpaired electrons with one from a hydrogen atom, the hydrocarbon ethane can be formed, CH_3-CH_3.

Can you think of a molecular structure with two carbon atoms in which all the electrons are paired, but which involves only four hydrogen atoms?

Further bonding between the two carbon atoms is possible by pairing two more of the valence electrons thus:

$$\cdot\overset{\times}{\underset{\cdot}{C}}\overset{\times}{\underset{\times}{C}}{^\times} \quad \text{or} \quad \cdot C = C^\times \qquad\qquad \underset{H}{\overset{H}{>}}C=C\underset{H}{\overset{H}{<}} \quad \text{or} \quad CH_2=CH_2$$
$$\text{ethylene}$$

Bonding with hydrogen atoms then produces the hydrocarbon ethylene. Bonds which involve one and two electron pairs are known as *single* and *double* bonds, respectively.

single and double bonds

See if you can write down the structure of a hydrocarbon containing two carbon atoms bound by a triple bond.

When three electron pairs are shared between two carbon atoms a triple bond results.

$$\cdot C\overset{\times}{\underset{\times}{\overset{\times}{\vdots}}}C^\times \quad \text{or} \quad \cdot C \equiv C^\times$$

Pairing the remaining two electrons with the electrons of two hydrogen atoms leads to the hydrocarbon, acetylene.

$$H-C \equiv C-H \quad \text{or} \quad CH \equiv CH$$

The question may come to mind at this point, 'can carbon pair all four of its electrons with those of another carbon atom to form a diatomic molecule?' The answer is no, and you will be in a better position to understand why a little later.

Double and triple bonds (multiple bonds) occur widely in the compounds of carbon. Here again, a tremendous variety of hydrocarbon structure is possible with a given number of carbon atoms. Refer to the completed structures I and IV. Both had the molecular formula C_6H_{12}. We can write other structures (e.g. X to XII) which have this molecular formula but involve a double bond.

I 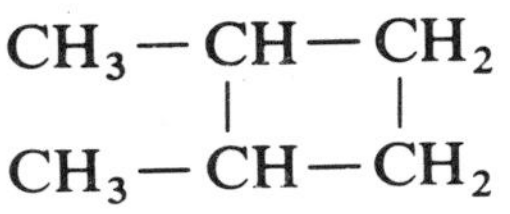 $CH_3-CH-CH_2$ $CH_3-CH_2-CH=CH-CH_2-CH_3$ X
          | |
       $CH_3-CH-CH_2$

IV CH_2 $CH_3-CH=CH-CH_2-CH_2-CH_3$ XI
      CH_2 CH_2
       | $CH-CH_3$
      CH_2
           CH_2 $CH_2=CH-CH_2-CH_2-CH_2-CH_3$ XII

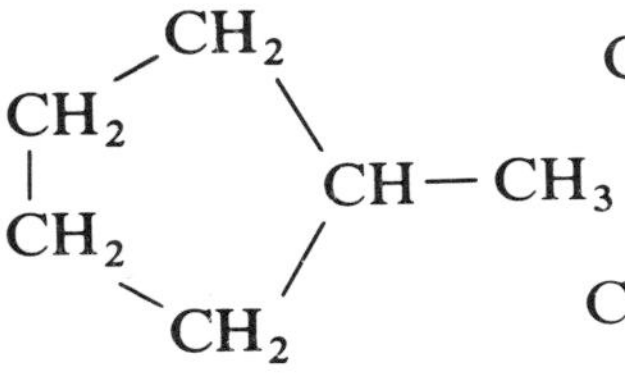

Write down the structures of all the possible isomers of the hydrocarbon C_4H_8 and C_4H_6. There are five and nine respectively.

Summarize what you consider to be the important points from section 10.2 and compare your summary with that given below.

See Answer 3, p. 94.

10.2.1   Summary of section 10.2

A carbon atom can form four covalent bonds to other elements. Carbon–carbon and carbon–hydrogen bonds are very strong and the compounds of carbon and hydrogen (hydrocarbons) are generally very stable. They can have a variety of structures including linear chains, branched chains and rings.

Structural isomers are compounds of the same molecular formula which have different structures.

Two, four and six electrons can be shared between carbon atoms to give single, double and triple bonds respectively.

Recommended parallel reading

The parts referred to in each book here, and at the ends of other sections, cover some but not necessarily all the subject matter of the section.

M. J. Sienko and R. A. Plane, *Chemistry: Principles and Properties*, McGraw-Hill, 1966, sections 3.3, 3.4 and 24.4 (the first two and a half pages).

Chemical Bond Approach Project, *Chemical Systems*, McGraw-Hill, 1964, section 7.16.

D. H. Andrews and R. J. Kokes, *Fundamental Chemistry*, Wiley, 1965, pp. 101–103.

S. Dunstan, *Principles of Chemistry*, Van Nostrand, 1968, pp. 63, 64, 104–107, 177–179, 183.

Nuffield Advanced Science Chemistry, Book I, Penguin Education, 1970, Topic 8, pp. 240–244, 249–252, 269–272.

SAQs relevant to section 10.2 are numbers 1 to 3.

10.3 The Uniqueness of Carbon

Although carbon compounds show all the characteristics typical of covalent compounds as a whole, the reverse is not true. No other element shows such diversity in the structures of its compounds. This is best illustrated by considering silicon, its nearest neighbour in the Periodic Table with the same number of electrons in its valence shell. (You learned in Unit 8 that the chemical characteristics of an element can be related to its position in the Periodic Table, so it seems reasonable to assume that all other elements will differ from carbon to a greater extent than silicon.) The fact that both elements lie in the same periodic group leads one to expect a degree of resemblance in the kind of compounds they will form. Thus, like carbon, silicon normally has a valency of four and forms reasonably stable bonds to other silicon atoms, carbon, hydrogen and many other elements. Some idea of the strength of these bonds relative to analogous bonds involving carbon may be obtained from Table 1 which shows so-called average bond energies. This is a parameter about which you need to know little at this stage except that comparisons between its values are believed to closely parallel those between the strengths or stability of bonds; the higher the figure, the more stable the bond. Significantly, the Si—Si bond is weaker than the C—C bond whereas the Si—O bond is stronger than the C—O bond. These values account for several differences in the chemistry of carbon and silicon. Thus, while carbon forms a great many compounds having linear chains, branched chains and rings of C—C bonds, silicon is less versatile.

Table 1

Average Bond Energies* of Some Important Bond Types in kJ mol^{-1}

C—H	415	Si—H	320
C—C	345	Si—Si	220
C—O	355	Si—O	450
C—N	305	F—F	150
C—F	485	Cl—Cl	240
C—Cl	335	Br—Br	190
C—Br	280	I—I	145
C—S	280	S—S	225
		N—N	160
		O—O	140
		Na—Na	70

Again, compounds of hydrogen and silicon are relatively unstable compared with the hydrocarbons and react avidly with oxygen to form silicon dioxide, which contains the very strong Si—O bonds. The simplest

* *These values are derived ultimately from quantitative experimental studies of certain chemical reactions. The energy of a particular bond-type will vary slightly from compound to compound. The figures quoted are average values. You do not have to remember them.*

hydrocarbon, methane (CH_4) is the main constituent of natural gas and its controlled combination with oxygen (i.e. burning) forms the basis of its use as a fuel.

$$CH_4 + 2O_2 \longrightarrow CO_2 + 2H_2O$$

needs heat
to initiate reaction

$$SiH_4 + 2O_2 \longrightarrow SiO_2 + 2H_2O$$

reaction spontaneous at
room temperature

Many aspects of the roles of carbon and silicon in nature should be clearer to you in the light of what you have just learned. Thus, with your previous knowledge that oxygen and silicon are the most abundant elements in the Earth's crust and present knowledge that the Si—O bond is very strong, it should come as no surprise that compounds involving silicon and oxygen account for practically all rocks, clays and soils. Silicon occurs in flint, many types of sand, and in quartz, as silicon dioxide, SiO_2.

Another difference which emerges from a study of the compounds of silicon and carbon is that silicon does not form multiple bonds. The reason for this is not clear, but it is consistent with the general observation that elements in the second row of the Periodic Table do not participate in multiple bonding as readily as first row elements.

We have already seen the diversity of structures which carbon can form with hydrogen. The strengths of the carbon carbon and carbon hydrogen bonds are among the important factors which account for the existence of a huge variety of chains and rings. This is further extended by the ability of carbon to form multiple bonds. However, carbon is unique not only in the strength of the bonds it forms with other carbon atoms and hydrogen atoms, but also in that its bonds to many other elements are quite strong too (Table 1). This particularly applies to the halogens, and to oxygen, sulphur, nitrogen and phosphorus. Moreover, carbon can form multiple bonds with some of these elements. All the bonds are covalent.

Thus, we can build up a variety of structures involving carbon and other elements in which each atom participates in a number of bonds equivalent to its valency number; that is, the number of electrons it needs to acquire the next noble gas electron configuration. Let us focus attention on five elements, including carbon.

Element	H	C	N	O	F
Number of bonds (valency)	1	4	3	2	1
Valence shell	●	×●×	×●×	×●×	×●×

Nitrogen, oxygen and fluorine, having 5, 6 and 7 electrons in their valency shell, require 3, 2 and 1 more electrons respectively to attain the stable electronic configuration of neon.

Let us now build up some simple molecules involving these five elements. This is done in stages in Figure 3. Notice in Figure 3a that oxygen has two *pairs* of electrons in its valence shell and these are retained in the molecule. Such pairs are known as unshared pairs, non-bonding pairs or lone pairs. We have represented them in the molecule, following an established practice, by two dots. They play a vital role in the chemistry of compounds containing oxygen.

non-bonding electron pairs

The same applies in Figure 3b and Figure 3c for molecules containing nitrogen and halogen atoms. The last formulae drawn in columns a, b and c represent the standard abbreviations for molecular structures.

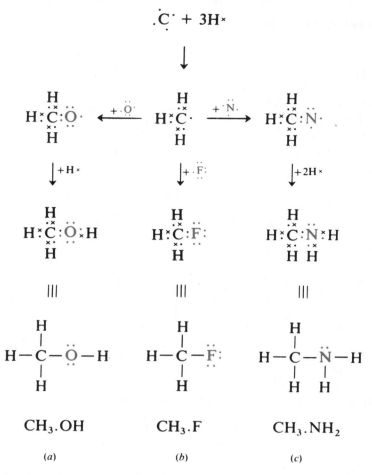

Figure 3 *Three simple carbon compounds and how they can be 'made up' from atoms of their constituent elements.*

Write down possible structural formulae in full and abbreviated form for compounds with molecular formulae C_2H_4O and C_2H_6NCl.

See Answer 4, p. 94.

Figure 4 shows a typical range of simple stable compounds involving these five elements.

Pick out compounds which are structural isomers.

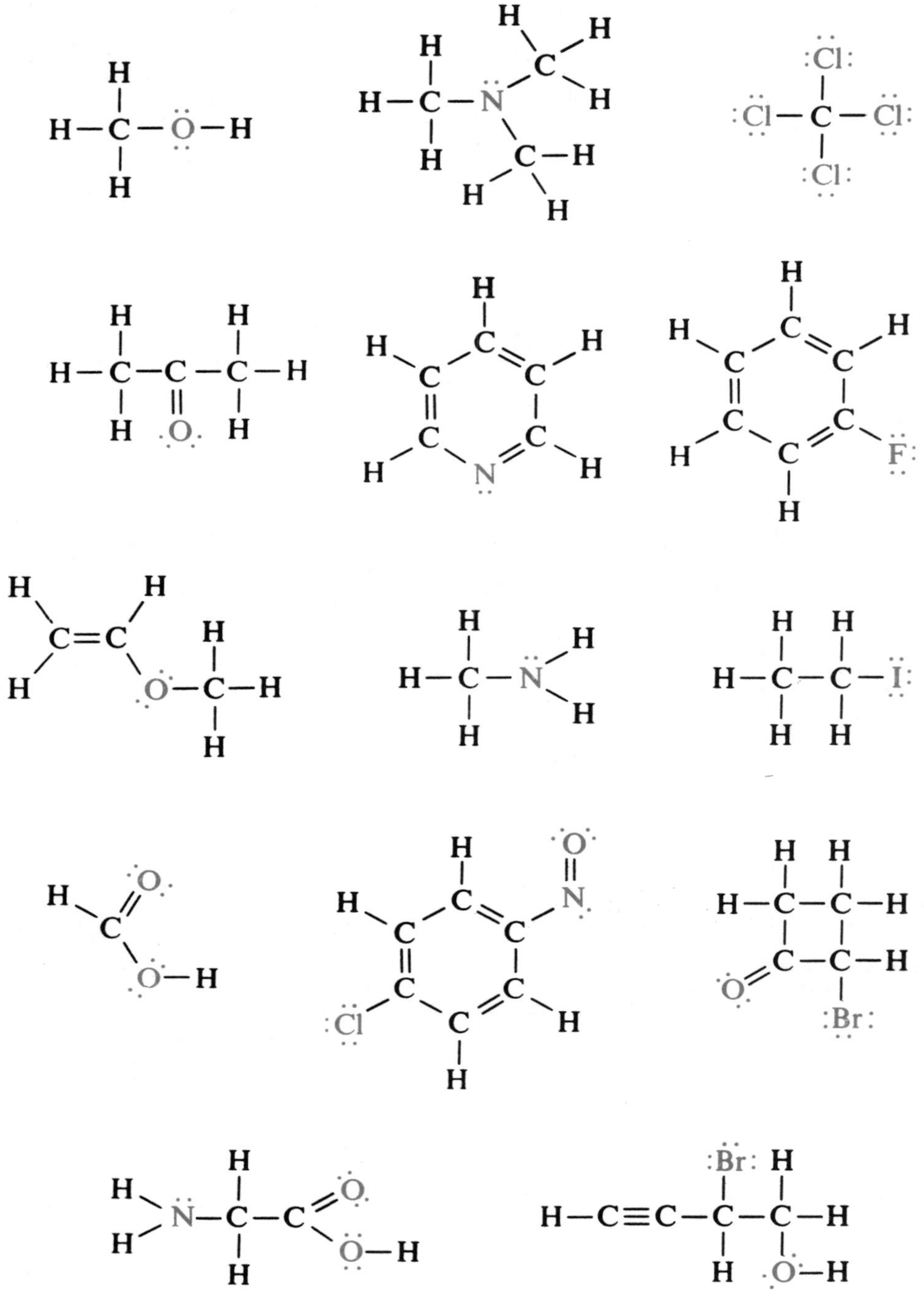

Figure 4 *Some typical stable compounds of carbon.*

The two pairs having molecular formulae C_3H_6O and C_4H_5OBr are structural isomers.

You may be interested at this stage to see the structures of some well-known compounds of carbon. These are given in Figure 5 (there is no need to memorize them!).

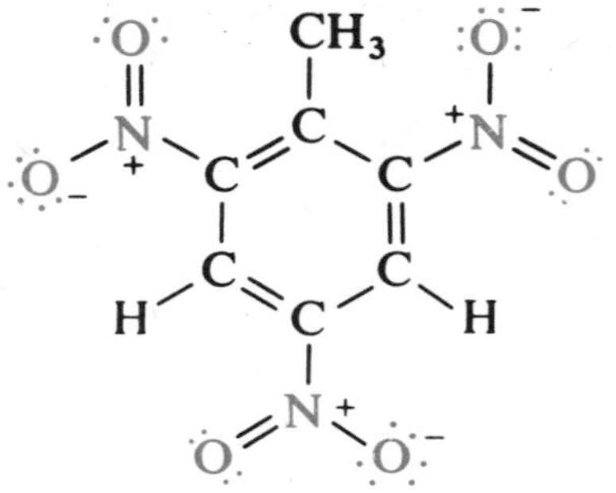

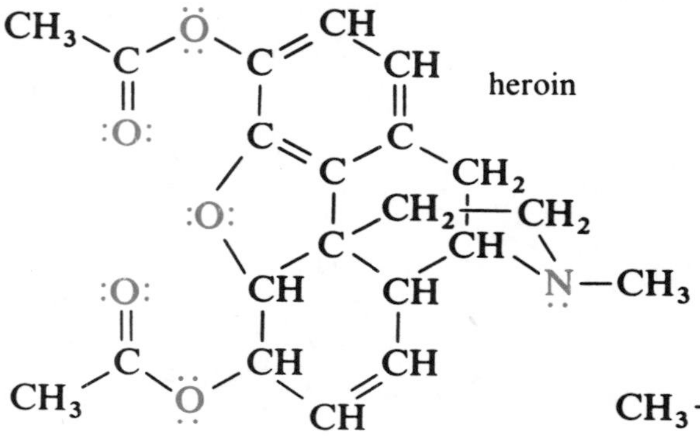

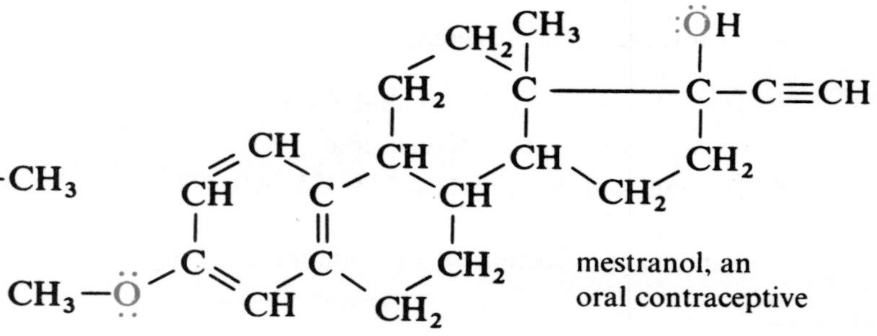

Figure 5 Two-dimensional representations of the molecular structure of some well known organic substances.

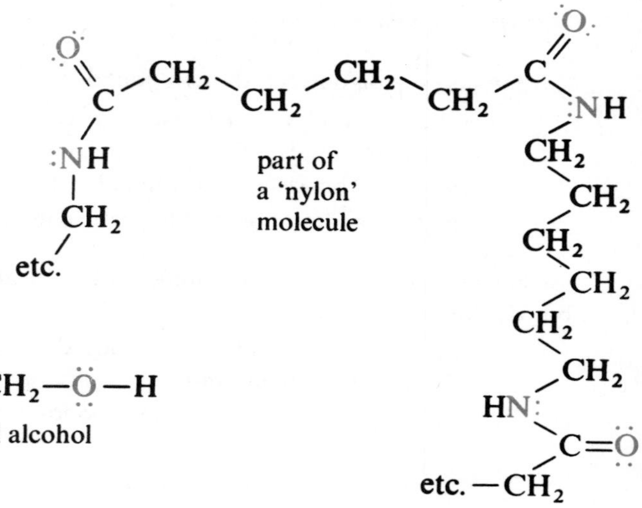

Diversity in molecular structure leads to diversity in chemical behaviour and no other element shows anything like the variety in its compounds that carbon does. It was this flexibility within the family of carbon compounds (combined with a favourable terrestrial environment) that led about three thousand million years ago to the development on this planet of a group of molecular species capable of self-replication. The evolution of compounds that were chemically capable of undertaking unique roles ultimately brought about life as we know it today (see later in Unit 21). The fact that carbon plays a key role in 'living structures' accounts for the use of the term *organic chemistry* to cover the chemistry of its compounds. Prior to 1828, all organic compounds were derived directly from living or dead organisms. In that year, the German chemist, Friedrich Wöhler, made the 'organic' compound urea (a chemical constituent of urine) from substances which could be obtained from non-organic (inorganic) matter. Thus, in a sense, he founded the detailed exploitation of organic chemistry that began with the dye industry in the nineteenth century.

organic chemistry

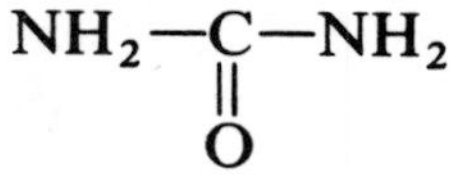

$$NH_2{-}C{-}NH_2$$

with O double bonded to C

urea

Today, pure carbon compounds synthesized in laboratories far out-number those that have been isolated from natural sources. This has come about in the present century because man has learned to exploit the variety in behaviour found in carbon compounds to the extent that he is now able to design (at the molecular level) substances which fulfil specific roles in his attempts to affect his environment.

Summarize what you consider to be the important points from section 10.3 and compare your summary with that given overleaf.

10.3.1 Summary of section 10.3

Carbon and silicon differ in the compounds they form because:

  (i) C—C and C—H bonds are more stable than Si—Si and Si—H bonds;

  (ii) Si—O bonds are exceptionally stable;

  (iii) Silicon does not readily participate in multiple bonds.

Carbon forms more covalent compounds than all other elements combined because:

  (i) stable C—C and C—H bonds allow the formation of a huge variety of chains and rings;

  (ii) carbon can form multiple bonds to itself and some other elements;

  (iii) carbon forms strong bonds with many other elements.

Recommended parallel reading

M. J. Sienko and R. A. Plane, ibid, section 24.5.

D. H. Andrews and R. J. Kokes, ibid., pp. 547, 687–689.

S. Dunstan, ibid, pp. 177–184, 373–375, 390–397.

*SAQ*s relevant to section 10.3 are numbers 4 to 8.

10.4 The Shapes of Organic Molecules*

Study comment

We have used many structural formulae in this section to illustrate the principles which govern molecular shape and electronic structure but you need only remember those mentioned in Objectives 14 and 18. However, you should attempt to grasp the principles themselves sufficiently well to achieve Objectives 11, 12, 13, 15 and 16.

10.4.1 The effects of electron repulsion

Electrons within a molecule occur in pairs, as far as possible, and each pair occupies a reasonably well-defined region of space, other electrons being virtually excluded from this space.† Hence, electron pairs repel one another and take up an arrangement in space that maximizes their distance apart at any given average distance from the nucleus. This simple principle predicts the approximate preferred arrangement of any given number of electron pairs in the valence shell of an atom in a molecule.

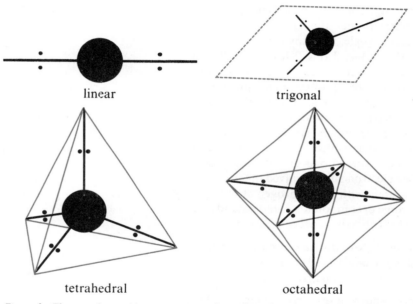
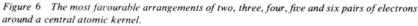
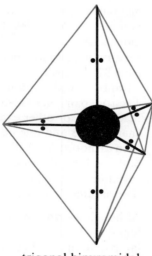
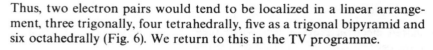
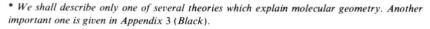

linear

trigonal

tetrahedral

octahedral

trigonal bipyramidal

Figure 6   The most favourable arrangements of two, three, four, five and six pairs of electrons around a central atomic kernel.

Thus, two electron pairs would tend to be localized in a linear arrangement, three trigonally, four tetrahedrally, five as a trigonal bipyramid and six octahedrally (Fig. 6). We return to this in the TV programme.

* *We shall describe only one of several theories which explain molecular geometry. Another important one is given in Appendix 3 (Black).*

† *We can explain this briefly in terms of the Pauli Exclusion Principle which some of you may have met in Unit 7 Appendix 1 (Black). You are not required to be able to reproduce this. The Pauli Exclusion Principle states that no two electrons bound to an atom can have the same four quantum numbers. All the electrons within one shell, the valence shell for instance, will have the same principal quantum number (n). Furthermore, it can be shown that the azimuthal (l) and magnetic (m) quantum numbers of electrons in a valence shell determine the directional characteristics of the covalent bonds which are formed when those electrons are shared with electrons of other atoms. Put in other terms, each region of space around an atom involved in covalent bonding has associated with it electrons with the same values of n, l and m. The Pauli Principle states that they must differ in their spin quantum numbers (s), which can only have one of two values ($\pm \frac{1}{2}$). Consequently, only electrons of opposite spin are allowed to occupy the same region of space, while electrons of the same spin must keep apart.*

Most covalent molecules are made up of atoms with four electron pairs in their valence shell. Methane, for example, is expected to be tetrahedral in shape, with the carbon atom occupying a central position and the four hydrogen atoms situated at the corners. In fact, the principle predicts that all carbon atoms within a molecule will be bonded to other atoms in this way. Furthermore, it leads us to expect that any atom surrounded by eight valence electrons (oxygen in water, for example) will have a tetra-hedral environment.

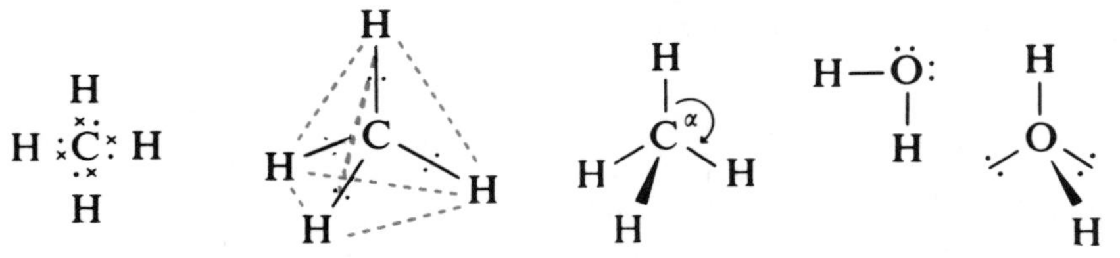

Note: The wedge-shaped symbol is a convention for representing a bond which points out of the plane of the paper, the thickest end being nearer to the observer.

Figure 7   The shapes of methane and water.

A tetrahedral configuration about an atom is characterized by a bond angle* of 109.5° (α in the structure drawn of methane, Figure 7).

It is now possible to measure bond angles in molecules experimentally, and such measurements have confirmed that the H—C—H angle in methane corresponds to the tetrahedral angle. In many other molecules, however, bond angles differ from this value. Table 2 lists ten different molecular structures, all but one of which contain a central atom with four electron pairs in its valence shell. Far from being 109.5°, bond angles vary from 92° for hydrogen sulphide to 180° for acetylene. It seems that a pure tetrahedral distribution of valence electrons is found only when all four electron pairs are 'identical'. The simple theory we have outlined is inadequate for other cases such as ammonia which has three bonding or shared pairs and one non-bonding or unshared pair of electrons. Again, each carbon atom of ethylene has two electron pairs involved in the double bond and the other two shared separately with two hydrogen atoms. The variation in bond angle within this range of compounds is understandable, if we assume that different types of electron pairs repel to different extents.

Although we do not have time to do so here (further discussion can be found in black-page Appendix 4), it is possible to deduce from experimental data of the kind given in Table 2 and to rationalize theoretically that:

(i) non-bonding electrons occupy more space than bonding electrons;

(ii) the four or six electrons of multiple bonds occupy more space than the two electrons of single bonds;

(iii) repulsion between shared electron pairs is less when they are drawn away from the central atom by an atom of electro-negative elements.

If one allows for the way these three factors distort structures away from the pure shapes in Figure 6, electron repulsion theory affords plausible explanations for most of the observed geometric arrangements of atoms within covalent molecules.

We are now in a position to consider in general the shapes (stereo-chemistry) of organic compounds. We shall need three dimensions for this, so have included in your Home Experiment Kit some ball and spring

* e.g. H—C—H bond angle is the angle subtended at the carbon atom by the two bonds between the carbon atom and hydrogen atoms.

Table 2

Some Bond Angles

a

		α
1. methane	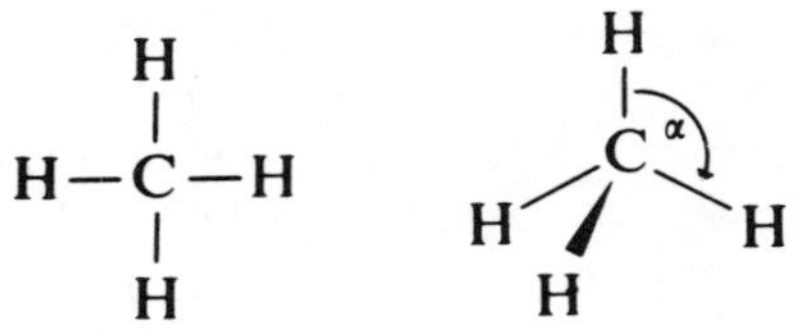	109.5°
2. ammonia	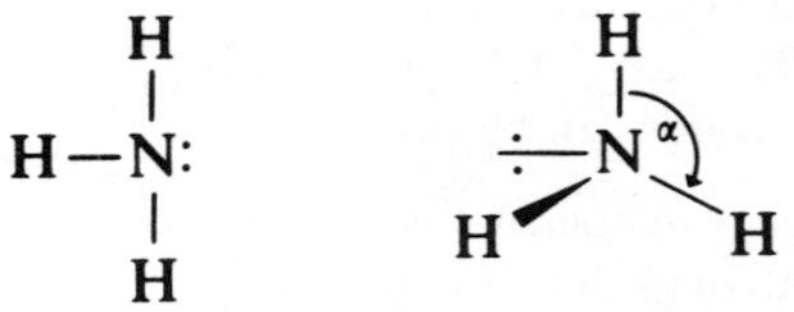	107.3°
3. water		104.5°
4. nitrogen trifluoride	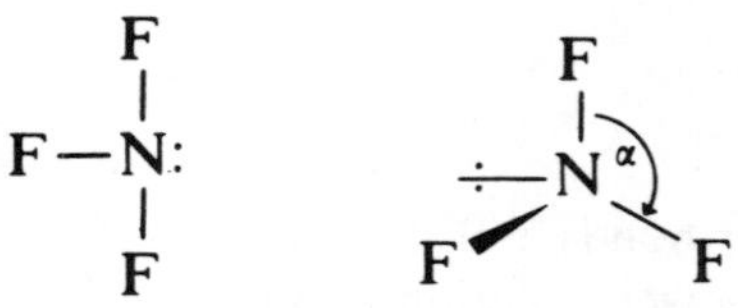	102.1°
5. hydrogen sulphide	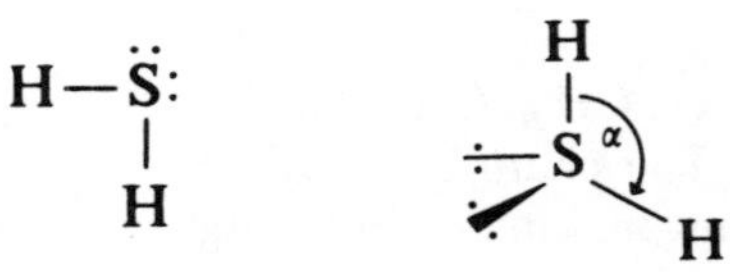	92°

b

6. boron trifluoride	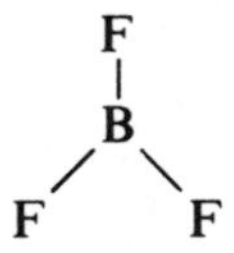	αFBF	120°
7. ethylene	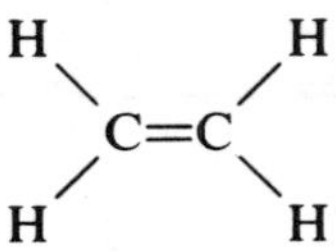	αHCH	116.8°
8. 1,1-difluoroethylene	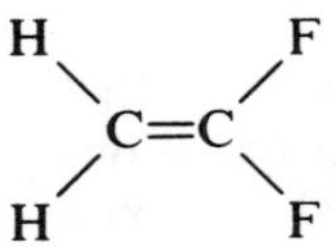	αFCF	109.3°
9. carbonyl fluoride	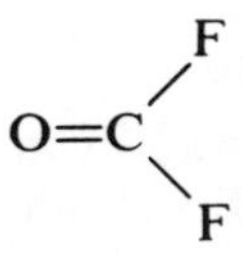	αFCF	108°
10. acetylene	H—C≡C—H	αCCH	180°

Do not attempt to memorize the data in this table.

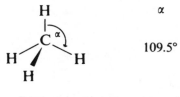

models. You will probably find these very helpful, but should be aware at the outset of their limitations. You should therefore read the white-page Appendix 1 on models at this point (p. 58), and refer to it again if in doubt about their limitations later on.

10.4.2 Molecules which contain single bonds only (saturated molecules)

Bond angles in non-cyclic (open chain) hydrocarbons are close to the tetrahedral value. Make a model of propane, $CH_3 . CH_2 . CH_3$. Notice that within the limitations of the model all single bonds can be rotated.

Although the carbon *framework* is restricted to a v-shape in this small molecule, the positions of the hydrogen atoms relative to those on another carbon atom can be varied considerably by C—C bond rotation. Draw the model in the space below, replacing the balls by the letter symbol of the appropriate atom and the springs by straight lines. Indicate on your drawing the bonds which allow rotation by a circular arrow as shown in the diagram below.

You may have drawn circular arrows on all your bonds and this would have been correct in the case of the model. However, rotation about a C—H bond is of doubtful significance as, unlike that about C—C bonds, we cannot recognize when it has occurred.

Now make a model of diethyl ether, $CH_3 . CH_2 . O . CH_2 . CH_3$. Notice again how flexible this open chain structure is. The chain as a whole can adopt a large number of overall shapes without making or breaking any bonds. Geometric arrangements of atoms within a molecule that can be interconverted by bond rotation are known as *conformations*.

conformations

> **We know that bonding electron pairs repel one another. Remembering that the springs represent bonding electrons, arrange the "backbone" of the model of diethyl ether in what you consider to be the most stable conformation.**

See Figure 8 on plate facing p. 96.

Now arrange the model in its least stable conformation, that is, that which resembles a 5-membered ring. Here, the bond–bond repulsions between the two terminal CH_3 groups are severe. So diethyl ether can adopt a range of conformations from the most stable, stretched-out, structure to the least stable, almost cyclic, structure.

Although the difference in stability between these two extremes is considerable, that between the conformations of intermediate stability is quite small and, at ordinary temperatures, the bonds of diethyl ether are rotating very rapidly, no single conformation having a lifetime of more than 10^{-13} seconds. Compare this with the cyclic molecule, tetrahydrofuran, which you can make from the 'pseudo-cyclic' conformation of diethyl ether by removing one C—H bond from each CH_3 group and joining the two carbon atoms by a carbon carbon bond. This molecule is relatively rigid and molecular motion is restricted to ring vibration of the kind which you can demonstrate by twisting the molecule as shown below.

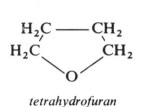

tetrahydrofuran

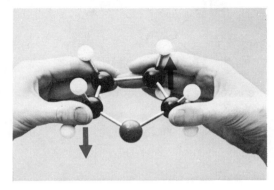

Figure 9 Relatively restricted conformational changes in tetrahydrofuran.

Thus, open chain compounds containing single bonds only (which for chemical reasons are referred to as 'saturated' molecules) are generally very flexible. Cyclic compounds are more rigid.

Would you expect rigidity to increase or decrease with increasing ring size? Find out by making a three-membered ring and an eight-membered ring. Use the six tetrahedrally drilled C-atoms and two O-atoms.

10.4.3 Molecules which contain multiple bonds (unsaturated molecules)

Carbon carbon double and triple bonds can be represented by ball and spring models by joining two tetrahedrally drilled carbon atoms with two and three springs respectively. Construct the molecules ethane, CH_3—CH_3; ethylene, CH_2=CH_2; and acetylene, $CH \equiv CH$. Then assuming for the moment that real molecules behave like the models, attempt to answer the following questions:

  (i) Which is the most flexible molecule?
  (ii) Which is the most rigid molecule?
  (iii) In which molecules is free rotation about the carbon carbon bond possible *and* meaningful?
  (iv) In which molecules do all the atoms lie in a single plane?
  (v) In which molecules are all the atoms in a straight line?
  (vi) Write down the molecules in order of decreasing carbon carbon bond length.
  (vii) Write down the molecules in order of anticipated increasing carbon carbon bond strength.

The sequence of correct answers is: (i) ethane; (ii) acetylene; (iii) ethane; (iv) ethylene, acetylene; (v) acetylene; (vi) ethane, ethylene, acetylene; (vii) ethane, ethylene, acetylene. In answering all these questions the models are true to life.* Thus, experimental measurements support the postulate that the introduction of multiple bonds (unsaturation) into a

---

* *However, as we used tetrahedrally drilled carbon atoms for our model, the bond angles in ethylene are not quite correct.*

molecule also introduces rigidity into it. Moreover, structures XIII and XIV represent recognizably different substances with different physical and chemical properties.

This shows they do not readily interconvert by rotation about the carbon carbon double bond.

Are they isomers?

XIII and XIV not only have the same molecular formula, which makes them isomers, but also the same bonding between the same groups within each molecule. They differ only in the spatial arrangement of these groups. Such a relationship is known as *stereoisomerism* and when it comes about, as in this case, because of restricted rotation, the two molecules are referred to as *geometric isomers*. (Geometric isomerism, then, is one form of stereoisomerism). By contrast with XIII and XIV, only one representation of the analogous single and triple bonded compounds XV and XVI can be recognized. Experimental measurements confirm that atoms involved in a carbon carbon double bond lie in one plane, XVII, and those involved in a triple bond are co-linear. Such measurements also show decreases in bond length with increasing multiplicity of bonding, to much the same extent as the models.

stereoisomerism

geometric isomers

The recorded bond energies (section 10.3) of C—C, C=C and C≡C in hydrocarbons are approximately 350, 610 and 840 kJ mol^{-1}.

You have learned enough to appreciate qualitatively the effects of multiple bonds on the two important parameters of molecular geometry, bond length and bond angle. Thus, in saturated systems we can expect bond angles involving carbon to be near 109° and, where double and triple bonds occur, the angles to be near 120° and 180° respectively. It is at this point that our models reach the limit of their usefulness, for although we can predict that departure from normal bond angles and bond lengths is going to lead to instability, we do not know how far one can go before the molecule is too unstable to exist. Table 3 may give you some idea.

> **Can you now give a reason for the negative answer we gave earlier to the question: 'Does carbon form diatomic molecules of the kind C ≡ C?'**

So much for the physical effects of multiple bonds on the structure. The three models of ethane, ethylene and acetylene you have made also enable us to gain insight into the effects of multiple bonds on chemical reactivity. Chemical reactions involve the making and breaking of bonds, that is, the

The location of four electron pairs between the two atoms would involve enormous interelectron repulsion. Actually, you may recall that the molecular spectroscopist in the TV programme for Unit 7 was, at one point, referring to the spectrum of the 'C$_2$ molecule'. He said that this molecule is so reactive that it only exists for a fraction of

reorganization of bonding electrons. The more accessible these are to outside reagents, the easier the reorganization becomes.

a second. Furthermore, other spectroscopic evidence confirms that the two carbon atoms are *not* connected by four electron pairs.

Arrange the three molecules we have just discussed in order of increasing chemical reactivity.

You should have placed ethane at the low and acetylene at the high reactivity end of the sequence. Thus in general, unsaturated compounds are expected to be more reactive than saturated compounds; a conclusion we would have also reached had we considered the relative electron pair repulsions within the three molecules. As we shall see later in the Unit, this turns out to be the case. The 'accessibility' of electrons, or their ability to respond to external effects, is termed *polarizability*. This is not to be confused with POLARIZATION, which was the term you met at the beginning of this Unit to describe an uneven distribution of electrons within a bond.

polarizability

Table 3

Structures which indicate the extent to which bond angles may be distorted.

stable at room temperature	unstable but known to exist	predicted to be very unstable and will probably never be isolated
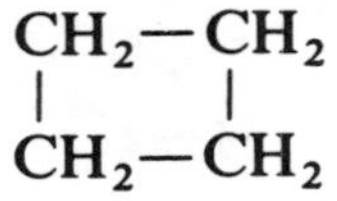 CH_2-CH_2 $\mid \quad\quad \mid$ CH_2-CH_2	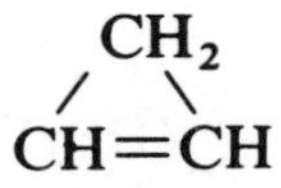 CH_2 $\diagup \ \diagdown$ $CH=CH$	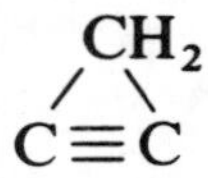 CH_2 $\diagup \ \diagdown$ $C\equiv C$
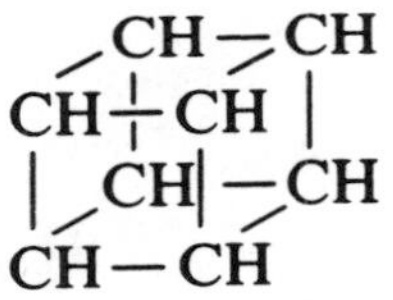 $CH-CH$ / $CH-CH$ / $CH-CH$ / $CH-CH$ (cubane-type cage)	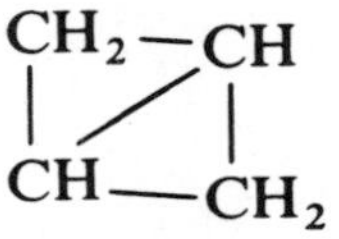 CH_2-CH $\mid \quad\quad \mid$ $CH-CH_2$	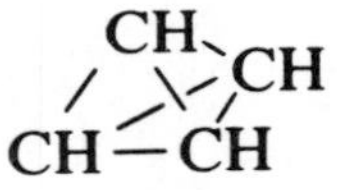 $CH-CH$ $CH-CH$
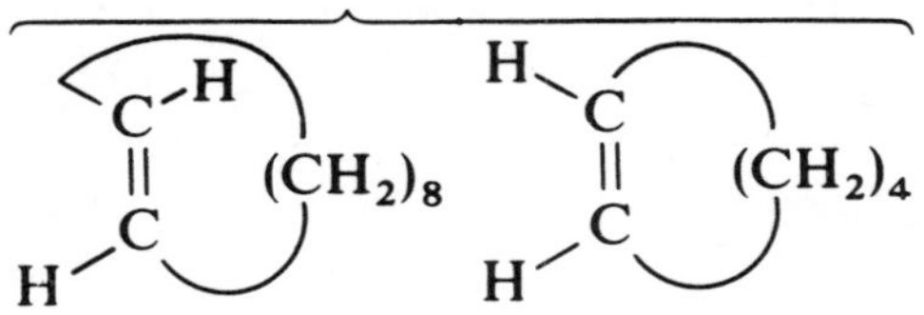 C^{-H} ‖ $C-H$ $(CH_2)_8$ and $H-C$ ‖ $C-H$ $(CH_2)_4$	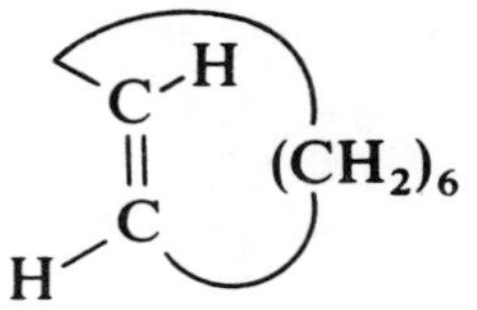 C^{-H} ‖ $H-C$ $(CH_2)_6$	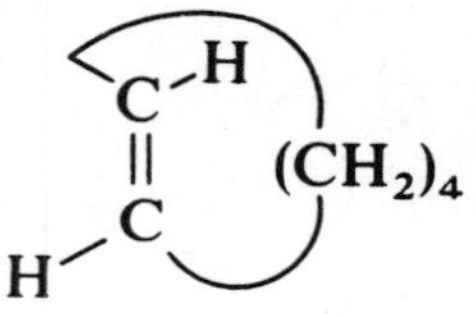 C^{-H} ‖ $H-C$ $(CH_2)_4$
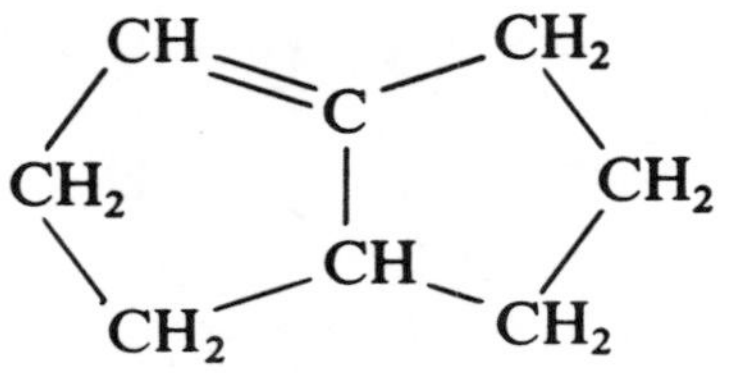 $CH=C$ ring with CH_2, CH_2, CH, CH_2, CH_2 groups	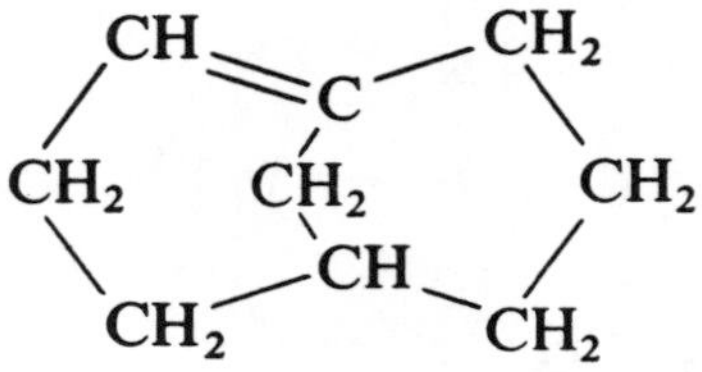 $CH=C$ ring with CH_2, CH_2, CH, CH_2, CH_2 groups	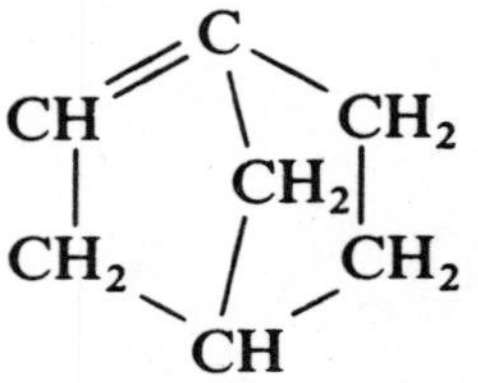 $CH=C$ ring with CH, CH_2, CH_2, CH_2, CH groups
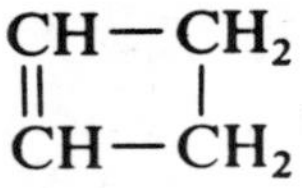 $CH-CH_2$ ‖ $\quad\quad \mid$ $CH-CH_2$	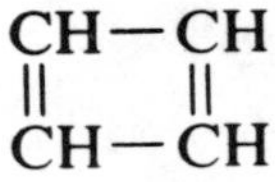 $CH-CH$ ‖ $\quad\quad$ ‖ $CH-CH$	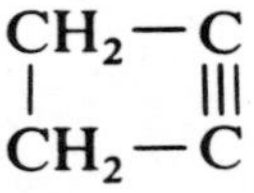 CH_2-C $\mid \quad\quad$ ⫿ CH_2-C

29

High polarizability usually means high reactivity. It so happens that most non-bonding electron pairs are also highly polarizable. This explains our earlier assertion that these play a vital role in the chemistry of organic compounds that contain such atoms as oxygen and nitrogen.

One type of structure, which is of great importance in organic chemistry and which might appear to contain double bonds, is, however, uncharacteristically inert. This is described in black-page Appendix 5, p. 63.

We now have most of the knowledge necessary for visualizing the three-dimensional structures of simple organic compounds. Three-dimensional structures of the compounds given in an earlier figure (Fig. 4) are shown in Figure 10, and those of more complex systems (Fig. 5) in Figure 11. If you have sufficient time you may care to construct some of these molecules, although you will not have enough pieces for all of them.

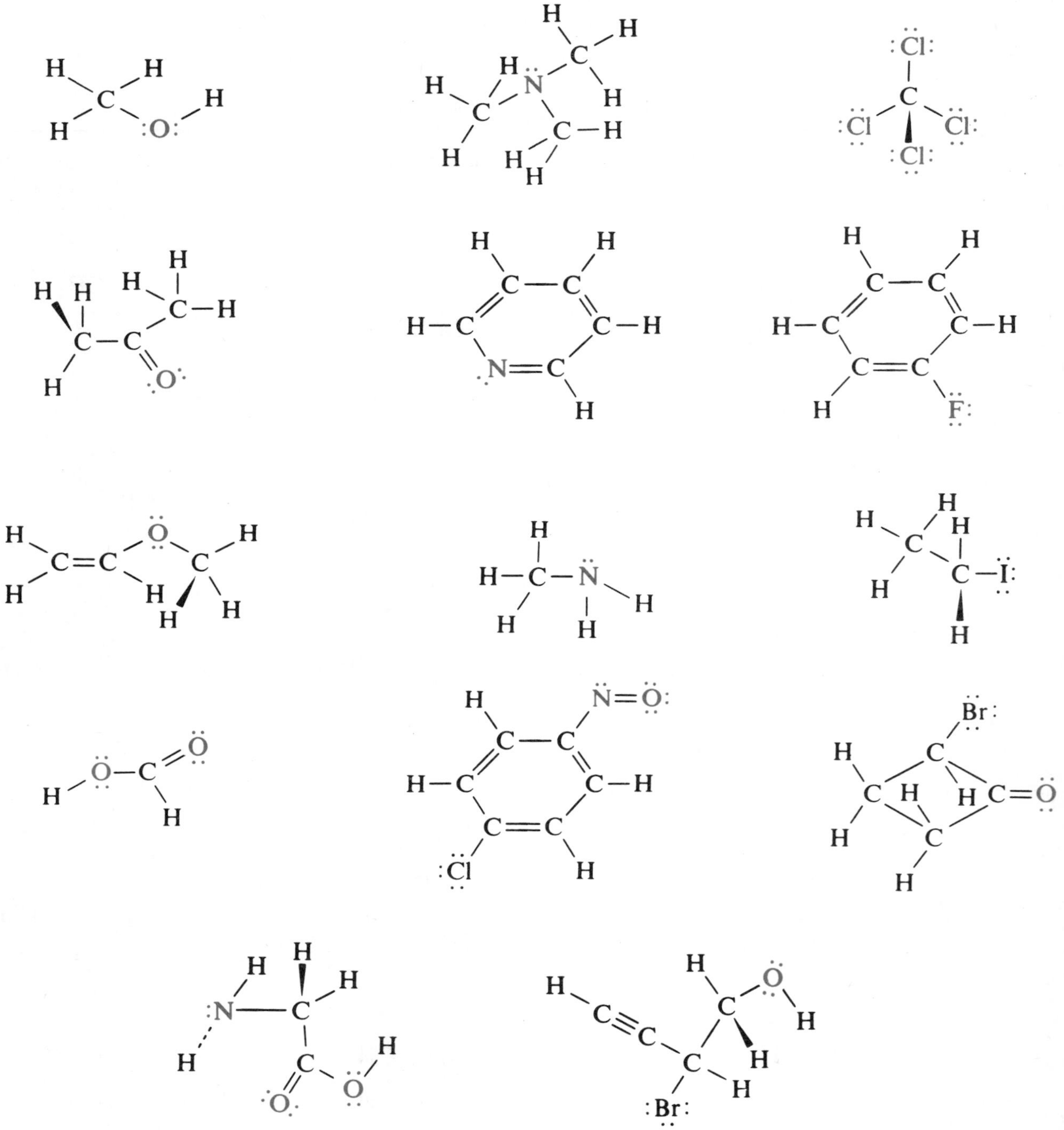

Figure 10   *Three-dimensional structures of the compounds given in Figure* 4.

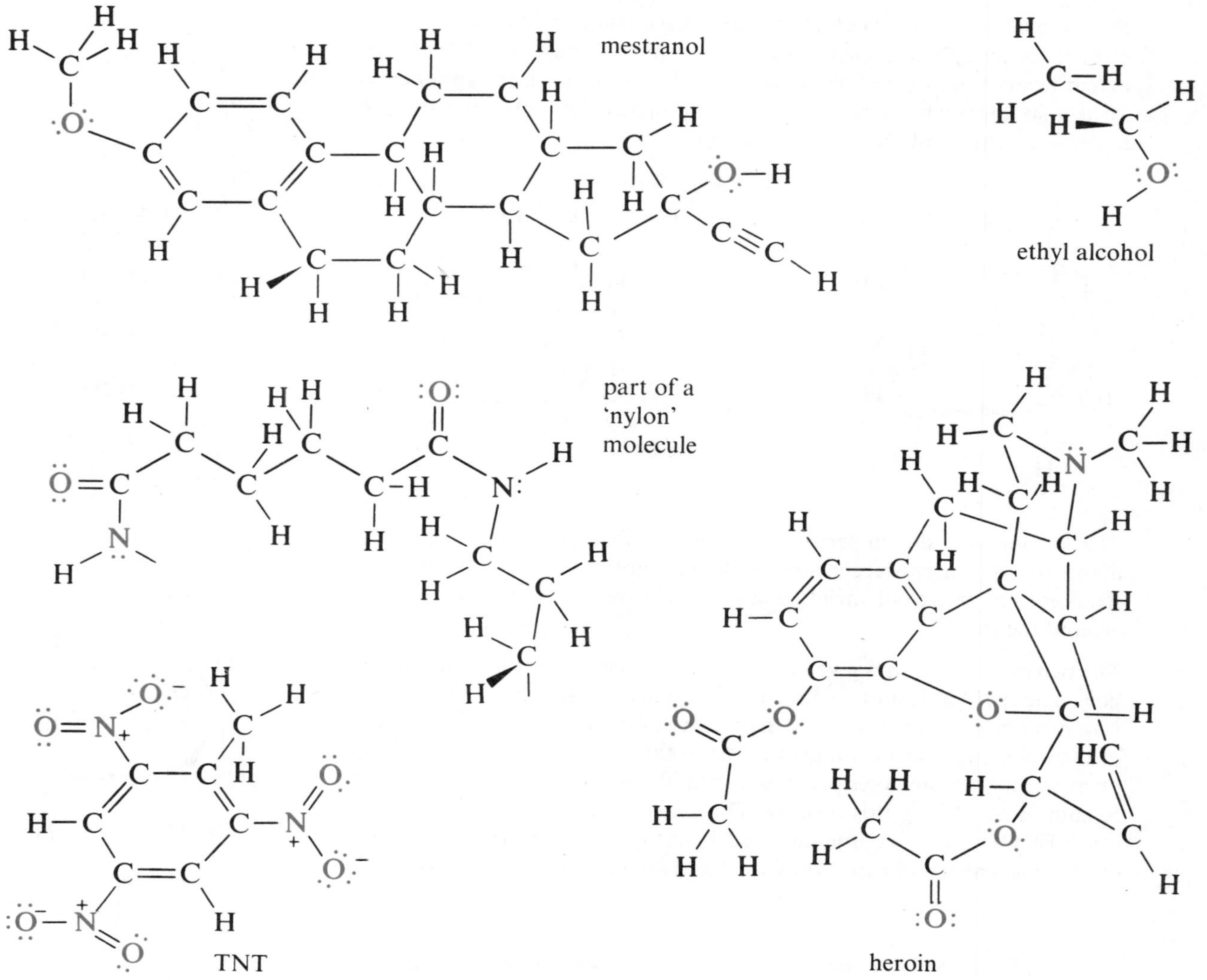

Figure 11   Three-dimensional structures of the compounds given in Figure 5.

10.4.4   Naturally occurring organic molecules

So far the principles governing molecular shape have been exemplified with simple organic molecules selected at random. We are going to end this section with a look at the structure of three types of compounds which occur in living organisms (often referred to as naturally occurring compounds or natural products).

All organic compounds were, by definition, naturally occurring until Wöhler's synthesis of urea, which we mentioned earlier. Today most pure compounds are manufactured, albeit from naturally occurring starting materials. Basic raw materials for organic compounds produced in bulk include oil, coal (both of which were formed from organic matter of living systems which existed tens of millions of years ago), wood, and animal and vegetable fat. Hundreds of other compounds made in smaller quantities are derived from these and many other sources. We cannot hope to mention in this Unit more than one or two types of naturally occurring organic compounds, so we shall focus attention on two which form part of the basic human diet, namely *carbohydrates* and *proteins*.

A. *Carbohydrates*
These compounds have the general formula $C_yH_{2y}O_y$ and are synthesized by plants from carbon dioxide and water. They include sugars, cellulose and starch. Most common sugars have six carbon atoms (the monosaccharides) and the most abundant and probably best known to you of

carbohydrates

31

these is glucose. Glucose can exist as an open chain or as one of two cyclic structures, though the open chain form is present only to a very small extent in solution. Five of the carbon atoms and one of the oxygen atoms form a six-membered ring to which are attached four OH groups, five hydrogen atoms and one CH_2OH group.

The formulation shown serves to emphasize that glucose is but one of many possible monosaccharides with this molecular formula and ring structure, the source of variety being the relative positions of the groups around the rings.

The two cyclic structures given differ in the relative positions of one H, OH pair. The carbon atom to which these groups are attached is where the ring can open to give the open chain form of the molecule. Ring closure of the open-chain form can give either of the two cyclic structures. Ring opening and closing takes place more readily in solution. All other bonds remain intact during this process. Thus, a structure in which one of the *other* H, OH pairs are interchanged is not glucose but another monosaccharide which is chemically distinct and fulfils a different role in nature.

mannose

Mannose is clearly isomeric with glucose. What kind of isomer? The answer is underneath the next question.

Consider the bond angles implied by the planar formulation of glucose given above and decide whether or not this is a realistic representation of the molecule's shape.

Glucose and mannose are stereoisomers.

A planar hexagon implies internal angles of 120°, whereas we know that the ring carbon atoms of glucose are tetrahedral. The true structure of glucose then must involve deviation from planarity to allow for the expected tetrahedral bond angle of about 109°. Construct the ring skeleton of glucose using five tetrahedrally drilled carbon atoms and one oxygen atom. It is important that during this exercise you should work the springs round in the holes so that you do not get restrictions to rotation which are properties of the ball and spring model and not of the real molecule (see Appendix I)! If such restrictions are present you may find that your model has one or more slightly bent 'bonds'. If this is the case you should manipulate the balls (without breaking the ring) until you reach a shape where all the springs are straight. Draw the shape (6 lines will do) of your model. You will notice that the shape which gives no distortion of the springs, i.e. no departure from the tetrahedral bond angles, is not planar and your model will have one of the two conformations shown in the margin at the top of p. 34 (Fig. 12).

Moreover, by 'playing' with the model a little you will notice that these shapes or conformations can be interconverted without breaking bonds, that is, by bond rotation. It so happens that in the complete molecule of glucose, the so-called 'chair' conformation is the more stable. Make sure you can obtain this shape with your model.

Chair conformation of glucose (see also Figure 14 on plate facing p. 96).

Now place large springs in all the vacant holes and see if you can deduce why the 'chair' is more stable than the 'boat'.
(Convert it to the boat form and compare the two).

The 'chair' conformation is the one which minimizes bonding electron pair repulsions. Look particularly at the inter-relationship of the free springs you have just inserted.

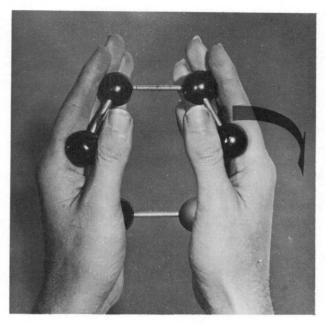

Figure 13   *Conversion of 'boat' into 'chair'.*

The bulk of carbohydrates in nature exist not as single monosaccharide molecules but as much larger molecules made up of chains of linked mono-saccharides (polysaccharides). For example, the paper which you have in front of you is largely the carbohydrate, cellulose, which can be shown, by breaking it down chemically, to consist of chains of glucose units joined together (Fig. 15). Some of these giant molecules contain over 10 000 glucose units. Cellulose occurs naturally as the main constituent of the fibrous material of practically every plant (including trees) and will be discussed at greater length in Units 13 and 14.

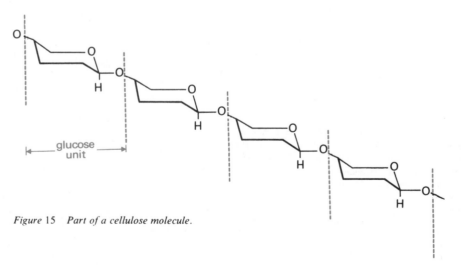

Figure 15   *Part of a cellulose molecule.*

B. *Proteins*

In much the same way that monosaccharides can be considered to be the building blocks of the large polysaccharides found in nature, amino acids are the basic units in the structures of proteins. There is an essential

proteins

difference, however. Polysaccharides are normally composed of one or two types of 'building blocks', whereas proteins consist of chains of up to twenty types of amino acids.

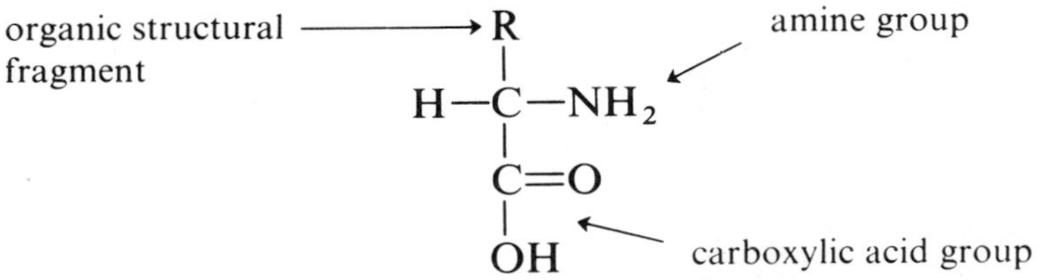

organic structural ———————→ R   amine group
fragment
$$H-\underset{\underset{\displaystyle OH}{\overset{\displaystyle |}{\underset{\displaystyle \|}{C=O}}}}{\overset{\displaystyle |}{C}}-NH_2$$ carboxylic acid group

General formula of an amino acid.

Protein structure is much more complicated than a simple chain of sub-units and this will be dealt with further in Unit 13. For our purposes, we need consider only the amino acids and the way they are linked in a protein. The structures of some commonly occurring amino acids are given in Figure 16. The links between amino acid units in a protein are formed between the amine group of one amino acid and the carboxylic acid group of the other.

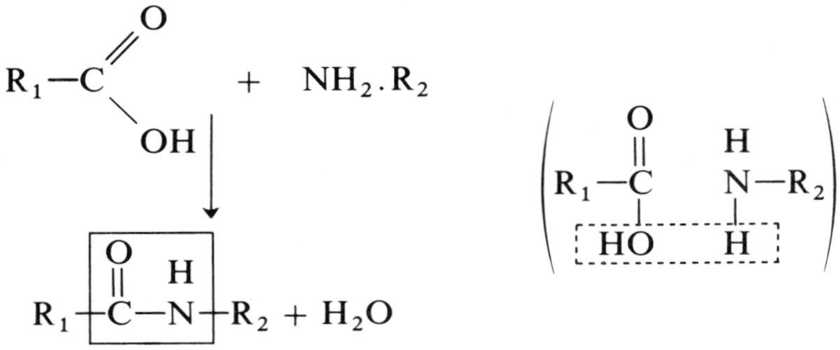

$$R_1-C\overset{\displaystyle O}{\underset{\displaystyle OH}{\big\|}} \quad + \quad NH_2.R_2$$

$$\downarrow$$

$$R_1\!-\!\!\boxed{\overset{\displaystyle O}{\underset{\displaystyle C}{\|}}-\overset{\displaystyle H}{\underset{\displaystyle N}{|}}}\!\!-R_2 + H_2O$$

$$\left(R_1-\overset{\displaystyle O}{\underset{\underset{\displaystyle HO}{|}}{C}}\qquad\overset{\displaystyle H}{\underset{\underset{\displaystyle H}{|}}{N}}-R_2\right)$$

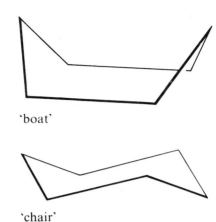

'boat'

'chair'

Figure 12   *Shapes of the ring skeleton of glucose.*

We only require that you remember the structures glycine and alanine.

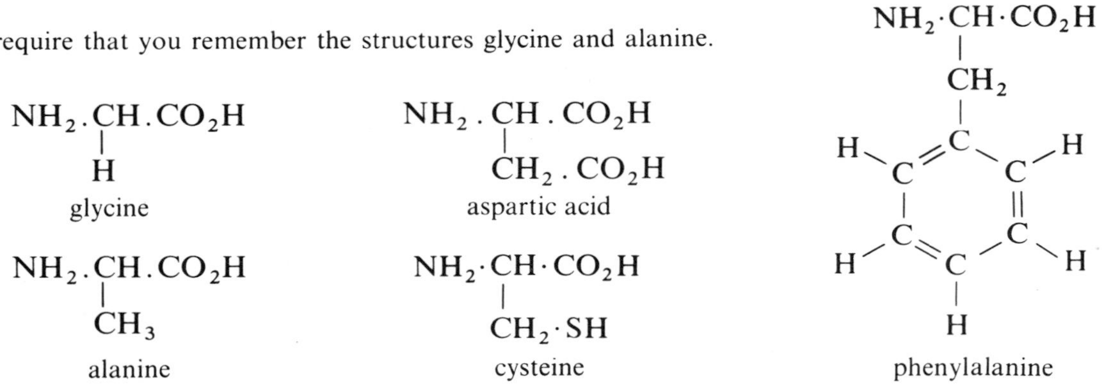

$$NH_2.CH.CO_2H$$
$$\overset{\displaystyle |}{H}$$
glycine

$$NH_2.CH.CO_2H$$
$$\overset{\displaystyle |}{CH_3}$$
alanine

$$NH_2.CH.CO_2H$$
$$\overset{\displaystyle |}{CH_2}.CO_2H$$
aspartic acid

$$NH_2\!\cdot\!CH\!\cdot\!CO_2H$$
$$\overset{\displaystyle |}{CH_2\!\cdot\!SH}$$
cysteine

$$NH_2\!\cdot\!CH\!\cdot\!CO_2H$$
$$\overset{\displaystyle |}{CH_2}$$
phenylalanine

Figure 16   *Some commonly occurring amino acids. (You need remember only the structures of glycine and alanine.)*

Write down the structure of the molecule formed when two molecules of glycine are combined in this way with the elimination of a molecule of water. Then see if you can write down the molecule which would be formed by condensing one molecule of glycine with one molecule of alanine. (Condensing, in this context, refers to the addition of one molecule to another with the simultaneous removal of a small molecule such as water).

See Answer 5, p. 95.

Check that you have drawn the correct structure and then write down the structure of the molecule formed by joining the following amino acids in the sequence given: glycine, alanine, glycine, cysteine, alanine, alanine, glycine, phenylalanine, aspartic acid.

What you have just written is typical of the fragment of a chain to be found in a protein. Such chains in proteins may consist of hundreds of amino acid molecules, each of which is chosen from a list of about twenty. The importance of the sequence in the chain will become apparent in later Units.

0.4.5 Chirality

Make a model of glycine, study it and then simplify it by replacing the entire $-COOH$ group by a red ball and the $-NH_2$ group by a blue ball. Place the molecule on the table and make another which corresponds to its reflection in an imaginary mirror standing to the right. Now take both models and try to superimpose each part on the other.

Is it possible?

The answer is yes. Now put the models back on the table in their mirror image relationship and replace the white ball (together with its short spring) which is closest to you in each model by a green one attached to a long spring. (The two new models should still be mirror image reflections of each other.) Having done this, try to superimpose one model on the other as you did before.

Is it possible?

The answer is no. Confirm that you have understood these instructions by looking at Figure 17, plate facing p. 96.

Is it possible to interconvert the two models without breaking bonds?

The answer is no.

If each ball represents a different organic group, would you say the two molecules before you are those of different substances?

The two compounds are different in the same sense that your left hand and right hand are different. They are in fact demonstrating the universal geometric property of 'handedness' known as *chirality** (pronounced kyrality). We shall explore this property more fully in this Unit's TV programme.

chirality

Chirality is of particular importance in organic chemistry. For instance, more often than not only one form of a naturally occurring chiral molecule exists. This has profound implications in biochemistry as will become apparent in later Units.

Can you suggest a definition of a chiral molecule from what you have just been doing?

A molecule is chiral if it cannot be superimposed on its mirror image.

There is, in fact, a more convenient way of establishing chirality in a molecular structure than trying physically to superimpose two mirror-image models. It is found that objects are chiral if they lack symmetry.

Although there are different kinds of symmetry to look for if one wants to be certain, by far the commonest is a *plane* of symmetry; that is, an imaginary plane that will divide the object into halves, either one of which is the mirror image of the other across the plane.

* *Some of you may have met this before as 'asymmetry' or 'disymmetry'. Chirality is a less ambiguous term and, incidentally, was used to describe this property before the others came into common usage.*

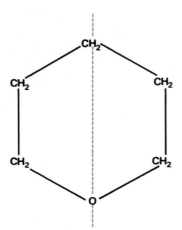

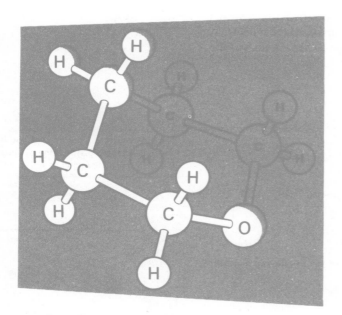

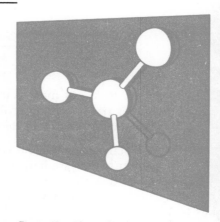

Figure 18   Planes of symmetry in two simple organic molecules.

We have demonstrated chirality with the two models in front of you. Take one of them and replace the green ball and its spring by a white one attached to a short spring. You demonstrated earlier that this model is not chiral (that is, it is *achiral*).

achiral

Can you see a plane of symmetry, that is, a plane which divides the model into two identical halves?

Satisfy yourself that no such plane exists in the other model. One source of chirality in an organic molecule, then, is a tetrahedral carbon atom which is attached to four different groups. A molecule containing such a unit will be chiral (provided it is not counterbalanced by another atom bound to the same four groups in the opposite sense!)

Study the list of amino-acids given in Figure 16 and decide which are chiral and which are achiral.

All but glycine are chiral. Let us take a closer look at the amino acid alanine. Convert the achiral model in front of you back into the other model's mirror image. These two models can represent the two forms of

Figure 19   *Plane of symmetry in a model of glycine.*

alanine where the white, green, red and blue balls represent H, CH$_3$, CO$_2$H and NH$_2$ respectively. We have predicted that they must represent two different substances. What experimental evidence is available to confirm that they are different? Each molecule contains identical atoms and identical bonding. Their reactivity towards all common chemical reagents is the same and they have the same melting point, boiling point and many other physical properties. The one unique property in which they differ is their interaction with plane-polarized light (explained in the white-page Appendix 2).

Historically, the study of chirality actually began with the observation that many compounds isolated from natural sources were able to rotate the plane of polarized light, and it was subsequently shown that the two forms of a chiral substance did this to the same extent but in opposite directions. They were thus termed *optical isomers* and chirality in molecules is referred to as *optical isomerism*. The molecules themselves are said to be optically active and each one of an isomeric pair is distinguished from the other by a prefix $(+)$ if it rotates the plane of polarized light to the right, and $(-)$ if it rotates it to the left. Glucose found in living organisms rotates to the right and so it is written $(+)$ glucose. The same applies to alanine, and the naturally occurring optical isomer is referred to as $(+)$ alanine.

optical isomers
optical isomerism

Optical activity would be lost if one form of a structure could be converted into its mirror image by the kind of bond rotation which we know generally takes place very easily in open chain single bonded systems; that is, if a molecule could by mere bond rotation assume a shape which contained a plane of symmetry. Generally, the two forms of an optically active molecule cannot be interconverted without bond rupture, although there are exceptions.

Of course, no net rotation would be observed if plane polarized light was directed through a solution of equivalent amounts of both optically active isomers.

This is the fourth type of isomerism you have met in this Unit. The others were structural, stereo and geometric. They are related in the following way.

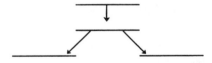

Insert one of the four kinds on each of the four horizontal lines to produce a 'family tree' consistent with their definitions.
(sections 10.2 and 10.4.3)

See Answer 6, p. 95.

It has been mentioned already that most chiral molecules made by living organisms are found in only one of the two possible forms.

Moreover, the unnatural form usually behaves differently from the natural form when introduced artificially into a biological environment. For instance, of the two glutamic acids, only the naturally occurring form $(+)$ functions as an effective flavour-enhancing agent for meats, soups and other foods. Again, only one of the chloramphenicol isomers is active against certain harmful bacteria. You may now have a little insight into the importance of chirality in protein structure, a subject which will be dealt with further in later Units. This whole question raises a fascinating point on which to speculate. How did it come about that we live in a world in which, for instance, both natural alanine and natural glucose are 'right-handed'; whereas for many other compounds, the naturally occurring form is the 'left-handed' one? If all the naturally occurring isomers

glutamic acid

chloramphenicol

(*See Note on page* 46).

37

of chiral molecules could be simultaneously inverted to the other form would we be any different? Certainly, optical isomers are chemically indistinguishable. We shall explore later the apparent contradiction that the two isomers of a chiral molecule can only be distinguished by interaction with polarized light, but that generally only one of a pair of isomers is of biochemical importance!

Summarize what you consider to be the important points from section 10.4 and compare your summary with that given below.

10.4.6 Summary of section 10.4

1. Electron pairs in molecules repel one another. Non-bonding electron pairs and those of multiple bonds repel other electrons more effectively than single-bonding electron pairs.

The spatial distribution of bonds around an atom in a molecule is found to be that which gives the minimum total repulsion between the electron pairs.

2. Free rotation about single bonds allows open-chain compounds to be very flexible. Cyclic compounds are less flexible and multiple bonds introduce rigidity into a structure.

3. Non-bonding electrons and those of multiple bonds are generally more *polarizable* (respond more readily to external effects) than those of single bonds.

High polarizability usually leads to chemical reactivity.

4. Glucose is a carbohydrate which exists largely as a cyclic structure. Six-membered rings involving tetrahedral carbon atoms (and in the case of glucose, one oxygen atom) are not planar but can exist in boat or chair forms.

5. Proteins are large molecules made up of sub-units derived from amino acids. The latter are linked together in a head to tail fashion by a process which involves the elimination of water.

6. All molecules fall into one of two geometric classes: they either can or cannot be superimposed on their mirror images; that is, they are achiral or chiral, respectively. Thus, a chiral molecule can exist as a 'right-handed' form or 'left-handed' form, both of which represent different substances. The two forms of a chiral compound are almost chemically and physically identical, although they can be distinguished by the way they rotate the plane of plane-polarized light. Practically all chiral molecules made by living organisms are found in only one of the two possible forms.

Recommended parallel reading

M. J. Sienko and R. A. Plane, ibid., section 24.4.
Nuffield Advanced Science—Chemistry Book I, Topic 8, pp. 242–244.
D. H. Andrews and R. J. Kokes, ibid., ch. 6.
S. Dunstan, ibid, pp. 78–87.

*SAQ*s relevant to section 10.4 are numbers 9 to 16.

10.5   The Influence of Structure on Chemical and Physical Properties

10.5.1   Introduction

In the present century, the correlation between properties of materials and their molecular structure has been recognized. This has not only increased man's understanding of the behaviour and properties of much of the material world around him, but has opened a new area of synthesis of new compounds designed to fulfil specific roles in the improvement of his environment. The rate of progress in this field has been astounding, so much so, that there is scarcely any field of human activity that has not been affected in some way by the design and synthesis of new materials by the chemist. Indeed, in some cases the human activity is itself a direct consequence of such advances. Before examining some examples, however, let us attempt to correlate some observable properties of organic materials with their structural features at the molecular level. It is useful at this point to sub-divide molecular structure into four components:

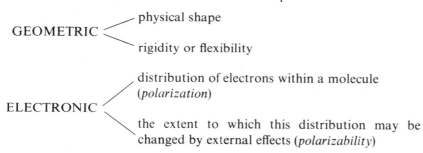

GEOMETRIC
- physical shape
- rigidity or flexibility

ELECTRONIC
- distribution of electrons within a molecule (*polarization*)
- the extent to which this distribution may be changed by external effects (*polarizability*)

(Most of the last section was concerned with the geometric aspect of molecular structure and you have met polarization and polarizability briefly earlier in the Unit.)

10.5.2   Melting point and boiling point sequences

You saw in Unit 5 that the melting point or boiling point of the substance (or, put in other terms, its existence as a solid, liquid or gas at a given temperature) is determined by the cohesive forces between molecules: the stronger the forces, the higher the melting point or boiling point and the greater the likelihood that the compound will exist as a solid or liquid rather than a gas at, say, room temperature.

Are there any particular molecular structural features which seem to influence the strength of the forces? Consider the series of compounds given in Table 4a.

Can you deduce one possible factor which correlates with the size of these forces?

Within a given series of structurally similar molecules, the cohesive forces increase with increasing molecular size. Indeed, the generalization can be

Table 4

Melting point and boiling point sequences (molecular weight in brackets).

a

	b.p.
CH_4 (16)	$-162°$
CH_3CH_3 (30)	$-88°$
$CH_3CH_2CH_3$ (44)	$-42°$
$CH_3CH_2CH_2CH_3$ (58)	$0°$
$CH_3CH_2CH_2CH_2CH_3$ (72)	$36°$

b

		m.p.
Na^+Cl^-	(58.5)	$801°$
$Li^+..Cl^-$	(42.5)	$613°$
$\overset{\delta+}{H_3Si}-\overset{\delta-}{Cl}$	(63.5)	$-118°$
$CH_3-\overset{\delta+}{CH_2}-\overset{\delta-}{Cl}$	(65.5)	$-139°$
$F-\underset{F}{\overset{F}{C}}-F$	(88)	$-185°$

c

Structure	m.p.		Structure		m.p.	b.p.
C_6H_5—C(=O)—OH (122)	121°		H—O—H	(18)	0°	100°
C_6H_5—C(=O)—O—CH_3 (136)	−12°		CH_3—O—H	(32)	−98°	65°
C_6H_5—C(=O)—O—C_4H_9 (178)	−22.4°		CH_3—O—CH_3	(46)	−138°	−25°
			CH_3—CH_2—O—H	(46)	−117°	79°
			CH_3—CH_2—O—CH_2—CH_3	(74)	−116°	35°

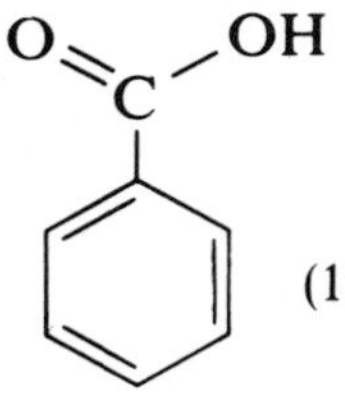

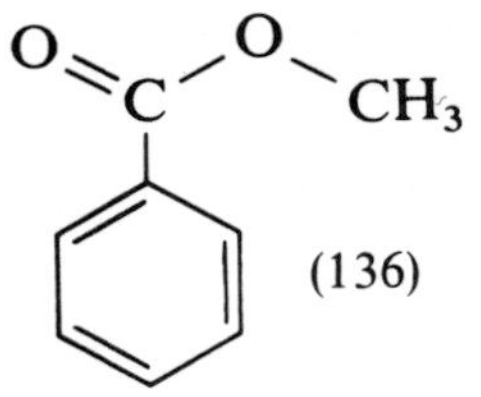

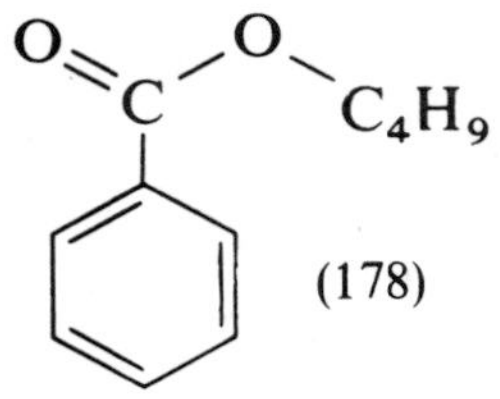

made that compounds with large molecules will have higher melting points and boiling points than those with small molecules.

Now look at Table 4b and name another facet of molecular structure which leads to high intermolecular cohesive forces.

Electrostatic attraction between oppositely charged ions in ionic compounds and between opposite charges in polar covalent molecules also increases cohesive forces. (You met this concept in Unit 8.)

Notice that, although the electronegativity difference between carbon and fluorine is considerable, the polarization of the carbon fluorine bonds in CF_4 is mutually cancelling and the result is a completely non-polar molecule and small intermolecular cohesive forces.

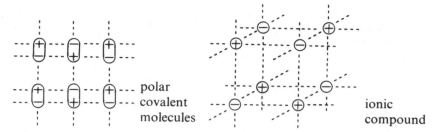

polar
covalent
molecules

ionic
compound

Figure 20   Cohesive forces in polar covalent and ionic compounds. (The dashed lines in this figure and the figure below represent attractive forces.)

Finally, look at the groups of compounds given in Table 4c. Bearing in mind that *the electronegativity of carbon is similar to that of hydrogen*, can you identify another structural feature which leads to large intermolecular forces?

Every molecule containing an O—H bond has a significantly higher melting point or boiling point than one similar in structure but without this bond. Note, for example, the melting points of ice (molecular weight 18) and $CH_3.CH_2.O.CH_2.CH_3$ (molecular weight 74) or the boiling points of $CH_3.O.CH_3$ and $CH_3.CH_2.OH$, which are exactly the same size. The difference is due to a phenomenon known as *hydrogen bonding*, in which OH-containing molecules are linked together by electrostatic attraction. You remember that water, for instance, has two non-bonding electron pairs. At the same time, it has slightly positive hydrogen atoms due to the polarization of the O—H bond. Molecules are thus attracted to one another in the way shown below.

hydrogen bonding

$$----\overset{\delta+}{H}-\overset{\delta-}{\underset{\underset{R}{\diagdown}}{\ddot{O}}}:-----\overset{\delta+}{H}-\overset{\delta-}{\underset{\underset{R}{\diagdown}}{\ddot{O}}}:-----\overset{\delta+}{H}-\overset{\delta-}{\underset{\underset{R}{\diagdown}}{\ddot{O}}}:---$$

Similar properties are observed for many covalent compounds containing hydrogen atoms bound to electronegative atoms particularly oxygen, nitrogen and fluorine. As you will see in later Units, hydrogen bonding is an important feature in the structures of biological giant molecules such as proteins and nucleic acids.

10.5.3   Solubility

When a chemical compound dissolves, the ions or molecules are separated, and the spaces in between are filled with molecules of the solvent. For this to happen, the cohesive forces between molecules in the original solid or liquid state of the compound must be overcome (cf. Unit 8).

The appropriate energy is supplied by attractive forces between the separate ions or molecules of the compound and the solvent molecules; that is, the old forces are replaced by new ones.

Figure 21   Solvation of an ionic compound by a polar solvent.

The very strong ionic bonds of an ionic compound must be replaced by strong 'solvating' bonds of a polar solvent (such as water), whereas the relatively weak cohesive forces in, say, a non-polar organic liquid (e.g. a hydrocarbon) can be replaced by similar intermolecular interaction with non-polar solvent molecules. In this sense, 'like dissolves like' is a very useful rule of thumb. Thus, water can dissolve sodium chloride, and hexane, $CH_3.CH_2.CH_2.CH_2.CH_2.CH_3$ can dissolve methane.

Can you think of a reason why methane will not dissolve in water?

Sodium chloride and methane, water and hexane, are of course, extreme examples. Many solvents have polarity between water and hexane and many polar covalent compounds will, in fact, dissolve to some extent in both water *and* hexane as well as solvents of intermediate polarity.

You saw in the last section that hydrogen bonding makes a powerful contribution to intermolecular forces. It stands to reason that it must also be important in solubility. In fact, it frequently dominates the characteristics of solutions and other liquids whose molecular structures involve —OH groups. Sugars (e.g. glucose) are covalent compounds. Although containing polar bonds, the molecules as a whole are not particularly polar. Despite this, they dissolve very readily in water.

Here the strongest cohesive forces are between the water molecules and these cannot be replaced by the limited interaction water molecules could have with methane molecules.

Remind yourself of the structure of glucose. Can you say why this sugar is readily soluble in water?

Glucose molecules contain lots of OH groups which can hydrogen bond very effectively with water.

10.5.4 Colour

The perception of colour involves a complex series of physiological and psychological responses to visible light (a small part of the electron-magnetic spectrum (Unit 2)) striking the retina of the eye. As you saw in Unit 6, it is a simple matter to demonstrate that white light is composed of radiation of various wavelengths, ranging from that corresponding to violet through blue, green, yellow and orange to red. When you observe a red object, it is absorbing all the wavelengths of visible light except red, which it reflects. Similarly, when you look through a piece of blue glass, the glass is absorbing all parts of visible light except blue. Materials show an infinite variety of tints and shades through selective absorption of various wavelengths within the visible region. Can we identify the molecular structural features of a compound which absorb light in the visible region and so produce colour? Table 5 gives a list of structural formulae of some coloured and colourless organic compounds.

See if you can deduce possible structural requirements for a compound to be coloured.

First of all, all the coloured compounds in the Table except XXVI have large numbers of multiple bonds. Furthermore, most of them, including XXVI, have atoms which are involved in a multiple bond and at the same time have non-bonding electrons.

Now you remember we said earlier that the electrons in multiple bonds are more 'loosely' bound or more polarizable than those in single bonds. Polarizability is greatly enhanced when multiple bonds alternate with single bonds. Such an alternating pattern is known as a *conjugated* system.

non-conjugated   non-conjugated   conjugated

42

Table 5

Structural formulae of some coloured and colourless organic compounds (c = colourless)

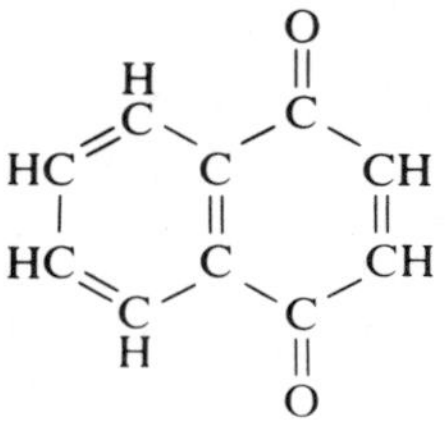

XIX yellow

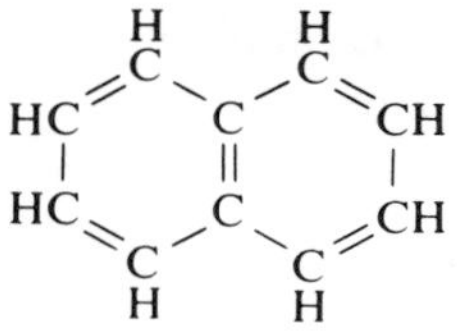

XX c

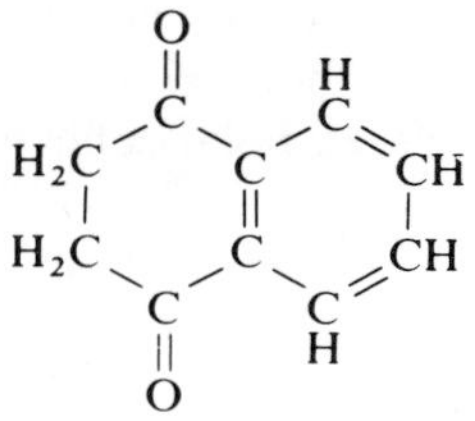

XXI c

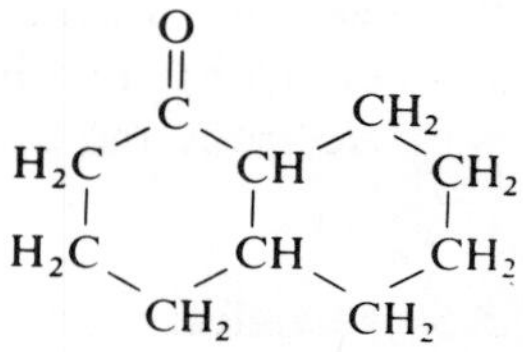

XXII c

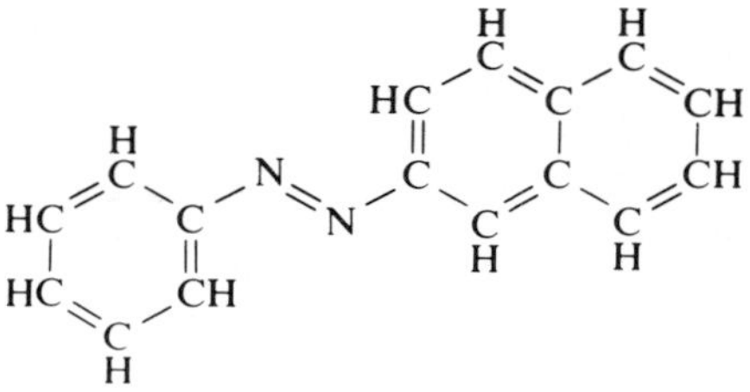

XIII orange

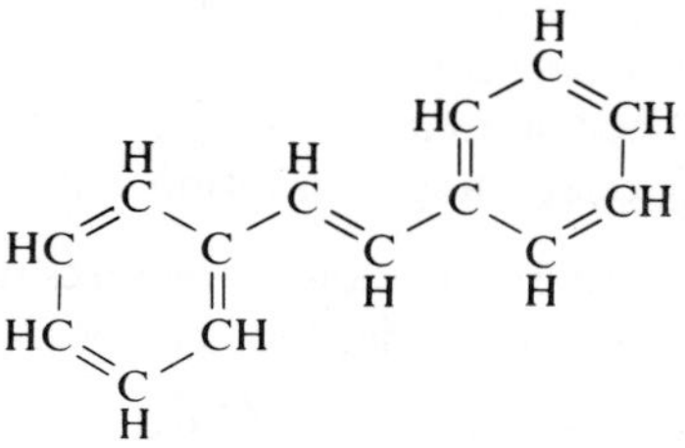

XXIV c

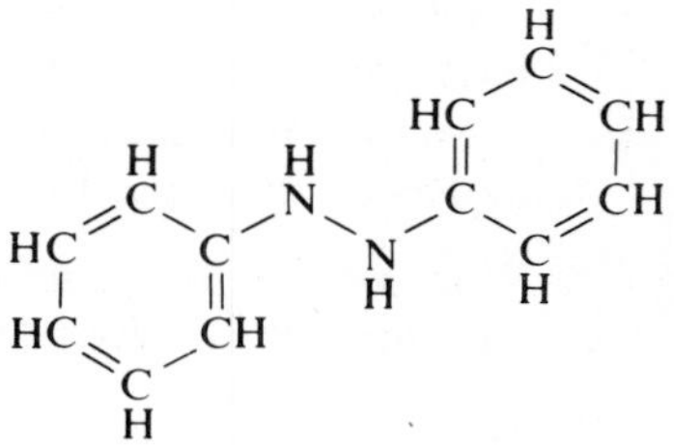

XXV c

$CH_2 = \overset{+}{N} = \overset{-}{N}$

XXVI yellow

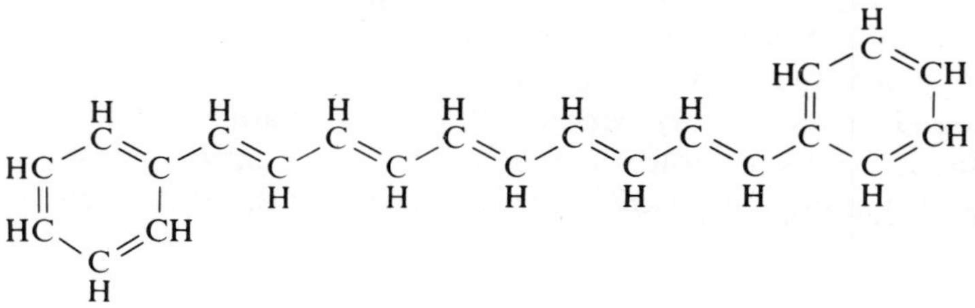

XXVII orange

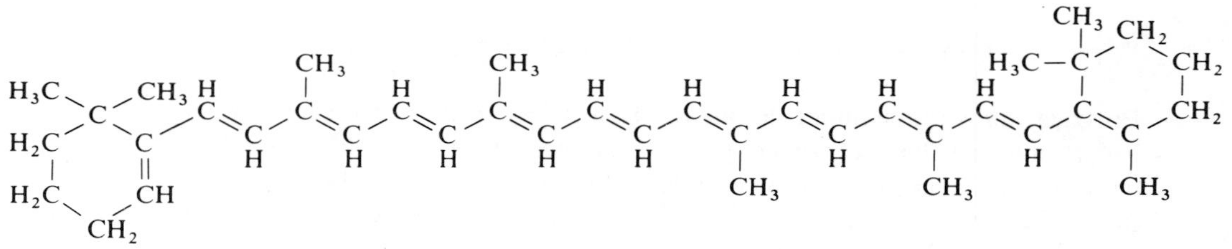

XXVIII red/orange

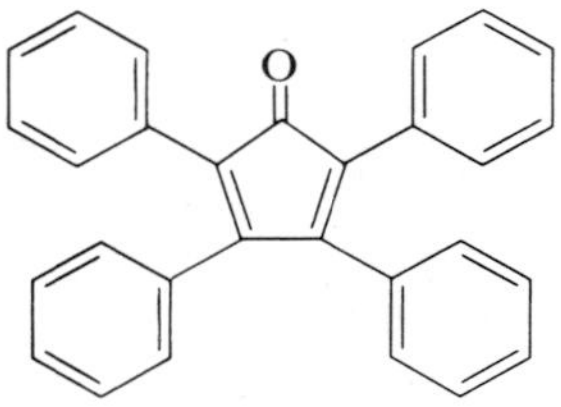

XXIX purple

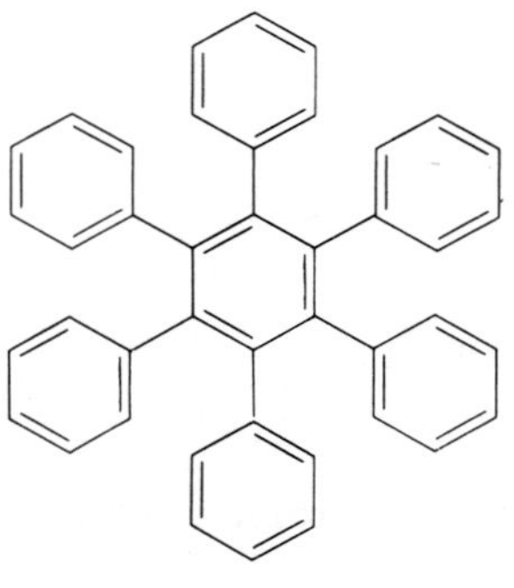

XXX c

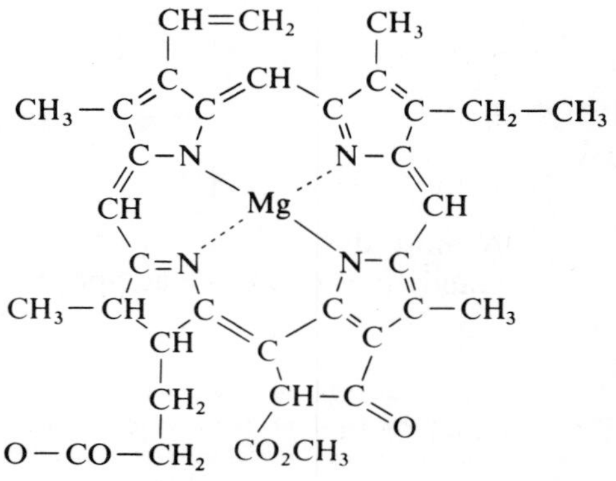

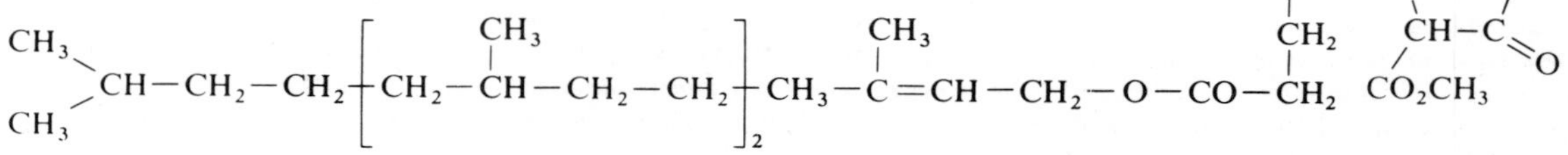

XXXI chlorophyll-a (green)

43

We also said that lone-pair electrons are highly polarizable.

It seems then, that absorption of visible light and, hence, appearance of colour in organic compounds occurs either: (i) in a very long conjugated system; or (ii) in a shorter one which involves atoms with non-bonding electrons; or (iii) in a multiple bonded system involving non-bonding electrons and charges. In all cases, polarizable electrons are involved.*

10.5.5 Chemical reactivity

An understanding of chemical reactions is of paramount importance in predicting correctly whether an organic compound is going to be stable or react in a particular way under certain circumstances. It is also implicit in the quest for knowledge of how one substance is converted to another within living organisms and, indeed, in man's attempts to synthesize new materials or naturally occurring ones whose extraction is uneconomic.

What we would like to do here is to briefly indicate features of molecular structure which determine the way a particular molecule is likely to react.

The chemical behaviour of a compound is dominated by its molecular *electronic* structure, although geometric structure can be very important in some cases. (We shall return to this shortly.) Chemical reactions involve the reorganization of electrons, so clearly, the more polarizable the electrons of a molecule, the more likely they are to become involved in bond making and breaking. Consequently the chemical reactions of a compound are frequently concerned with the multiple bonds or non-bonding electrons its molecules contain, and the relative inertness of the saturated hydrocarbons (which contain neither of these) can therefore be understood.

$$CH_3-CH_3 + HBr \longrightarrow \text{No reaction}$$

$$CH_3-NH_2 + HBr \longrightarrow CH_3-\overset{+}{N}H_3Br^-$$

$$CH_2{=}CH_2 + HBr \longrightarrow CH_3-CH_2-Br$$

Polarization within a molecule can also be a source of chemical reactivity. Polar molecules have centres of positive and negative charge, which tend to attract the oppositely charged parts of other polar molecules, and, in cases where energetic factors allow (Units 11 and 12), chemical reactions may ensue:

A more detailed discussion of the influence of electronic structure on reactivity is given in black-page Appendix 6.

* *Practically all the colour you observe is connected with structural features of this kind except that involving compounds of certain metals where the electronic structure of the metal atom determines the absorption characteristic of the material. There is another source of colour in our environment not connected with absorption but with scattering of light. Can you name it? See the answer at the foot of page 46.*

The geometric structure of a molecule becomes important in cases where the reactive site is physically shielded by other groups attached to the same structure, and in such situations chemical reaction may not occur at all to any measurable extent.

A much more profound influence of shape on chemical reaction is encountered in reactions which take place in living organisms. These are brought about with the assistance of large protein molecules known as enzymes. (You will learn more about them in later Units). The function of the enzyme is to bind itself to the molecule undergoing change, and thereby allow it to be transformed along a pathway which would have otherwise been extremely difficult.

$$A \nrightarrow B$$ (energetically impossible in biological environment)

$$A + En \longrightarrow EnA$$
$$EnA \longrightarrow B + En$$ (energetically possible)

The actual process is much more complex than that indicated and will be discussed again in Unit 15. Yet this simple picture is adequate to illustrate the point we are about to make. Like most naturally occurring molecules, enzymes are chiral and will only accept one optical isomer of a substance. You can demonstrate this with your model. Make the two mirror image tetrahedra you had before. You will find that only one of them will stand on Figure 22 so that the coloured balls match those given in the circles. Figure 22, then, represents the chiral active site of the enzyme and will not 'accept' the other isomer. Enzymes themselves are made by processes involving, ultimately, other enzymes and so the production of only one of a pair of optical isomers is self-perpetuating.

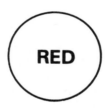

Figure 22   *A chiral template.*

Summarize what you consider to be the important points from section 5 and compare your summary with that given below.

10.5.6 Summary of Section 10.5

Factors which influence melting points and boiling points

   (i) They increase with increasing molecular size.

   (ii) They increase with increasing charge separation (ionic character) in a series of compounds.

   (iii) They are higher than would have been expected from (i) and (ii) in covalent compounds containing hydrogen atoms bound to oxygen, nitrogen and fluorine. This is due to the phenomenon of hydrogen bonding (illustrated for water below).

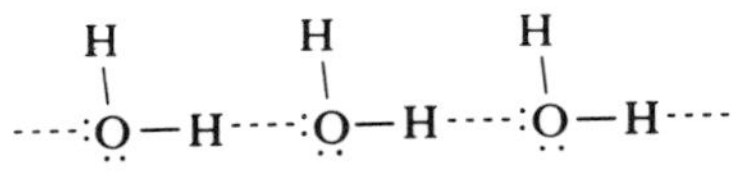

Factors which influence solubility

The ability of a liquid to dissolve a compound is determined by its ability to replace the intermolecular attractive forces in the solid state by similar forces in solution. This means that 'like dissolves like'.

Factors which determine colour

The presence in the molecule of one or more of the following:

   (i) a very long conjugated system of double bonds;

   (ii) a short conjugated system of multiple bonds which involves atoms with non-bonding electrons;

   (iii) a multiple bonded system involving non-bonding electrons and charges.

Factors which determine chemical reactivity

Chemical reactions take place most readily with compounds whose molecules contain polarizable electrons, i.e. compounds with multiple bonds and/or non-bonding electron pairs.

Polarization within a molecule frequently leads to chemical reactivity. Although the dominant structural factor determining chemical reactivity is the electronic structure of a molecule, the geometric structure can be important in certain cases, and is so for virtually all the chemical reactions that occur in living organisms; that is, those that take place with the assistance of enzymes.

Recommended Parallel Reading

S. Dunstan, ibid., pp. 113–126, 153–164.

D. H. Andrews and R. J. Kokes, ibid., pp. 270–272; ch. 26.

SAQs relevant to Section 5 are numbers 17 to 22.

Answer to question in the footnote on p. 44.
The blue of the sky, for example.

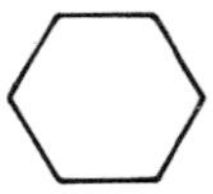

Additional Note (refers to page 37):
The symbol A (see margin) is shorthand for benzene (B). The replacement of a hydrogen atom by a bond allows the general group to be incorporated into a larger structure.

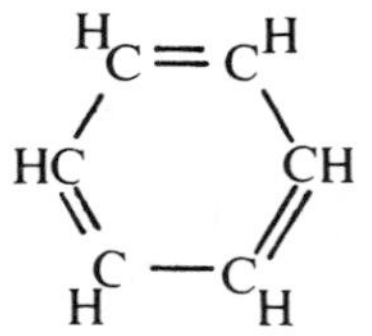

A

B

10.6 The Influence of Structure on Properties — Examples from Chemical Technology

Study comment

This is largely a descriptive section which refers in many places to specific chemical compounds and their molecular structures. You are not required to remember their names or precise structures, though you are expected to recognize structure types associated with each of the industries mentioned. Where relevant, you should also be able to relate the way a class of compounds is utilized to its structure type (Objective 22) and achieve Objective 23.

10.6.1 Motives and methods of chemical technology

You will probably find it useful whilst reading section 10.6 to keep in mind the broad context of each topic within the field of organic chemical technology. You can do this by referring to the flow sheet of Figure 23 (do not worry about box X for the moment — we shall refer to this at the end of the Unit).

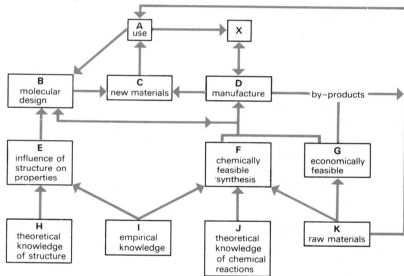

Figure 23

We are particularly interested in the sequence HEBCA but let us start by taking an overall view of motives and methods. Two kinds of motivation for the production of new materials can be recognized: firstly, a requirement for a substance with particular properties (AC, BD, etc.); and, secondly, the appearance of new resources and the desire to use them (AK). The most remarkable example of the latter is the petroleum industry, which now provides the bulk of the raw materials used in the manufacture of organic substances. Crude oil, a complex mixture of hydrocarbons with small amounts of nitrogen- and sulphur-containing organic substances, was initially exploited only as a source of fuel. But it soon became clear to the chemists in that industry that petroleum chemicals could be made to undergo economic chemical transformations into a versatile range of products with wide application in modern society.

The first motive, however, is likely to become increasingly dominant as growing chemical knowledge enables substances with particular required properties to be synthesized at need. Once the desired properties of a new material are identified (C, AB) its molecular design has to be considered, and again, although knowledge of the properties of other organic compounds (I) will always play an important role, increasing emphasis is to be expected on expanding theoretical knowledge of the structure/property relationship (H). Clearly, the feasibility of manufacture of the material has to be borne in mind throughout the design stage (BD). At the relatively small scale of laboratory preparation, one is usually concerned only with the chemical feasibility (F) of the synthesis of a new compound. On the industrial scale, however, economic considerations (G) are of critical importance and much weight is placed on, among other things, the availability of cheap raw materials (K) and possible economic use of any by-products.

As with design, the chemical feasibility of synthesis has relied in the past almost entirely on empirical knowledge; that is, chemical transformations analogous to those known for other compounds have been applied to new materials. Indeed, much of the synthetic work of today draws on the vast amount of empirical information accumulated over the past 100 years, although in most cases it is now understood why the chemical processes occur and thereby chemists have been able to effect improvements. Knowledge of chemical reactivity, however, has reached a point where for several years now chemists have been able to devise entirely new synthetic methods and this in turn has released more potential for man-made materials. In fact, this potential can be regarded as a significant bonus in the study of chemical reactions as opposed to chemical structure. Now let us consider briefly a few areas of the modern chemical industry in the light of this preamble.

10.6.2 Dyes

In addition to having the desired colour, a dye must adhere firmly to the fabric and must be chemically stable in the conditions likely to be encountered by the dyed object. The colour aspect of dye manufacture has largely developed empirically. Indigo (blue) and alizarin (orange-red) were used long before their structures were known. Many dyes used nowadays are based on the anthraquinone group found in alizarin.

anthraquinone

alizarin

indigo

the azo-group

The azo-dyes, as they are called, were developed from a chance observation in the last century. Dyes based on this kind of structure, e.g. direct black-38, are those manufactured in the largest quantity today.

The application of theoretical concepts to the molecular design of dyes has centred mainly on the aspect of adhesion. Here, both the dye structure and the structure of the fabric have to be considered at the molecular level. For example, material with polar groups such as wool can hold a polar dye molecule by electrostatic attraction and, in most cases, hydrogen bonding. Polar dyes, however, would not by themselves be suitable for the direct dying of many other relatively non-polar natural and synthetic fibres, such as polypropylene.

Martius yellow—a suitable dye for wool

polar and hydrogen bonding groups

a typical molecular fragment of wool

a molecular fragment of polypropylene

direct black-38

Polarity can be introduced into the fabric by incorporating metal ions in the material during manufacture or by the application of metallic compounds to the surface prior to dyeing. A suitable approach to the dyeing of man-made fibres is to disperse the dye in the material before it is spun and weaved.

It is fascinating to realize that there were no synthetic dye stuffs a hundred years ago. Clothes and most other man-made objects were a lot less colourful then than they are now, for the majority of pigments extracted from plants were browns, greens and dull reds, rather than the brilliant colours we have today. Since then, the development of organic chemistry has led to the production of thousands of colouring matters from inexpensive starting materials. In a way, the development of dyes founded the modern chemical industry and you will find this discussed in the historical booklet which accompanies the course.

10.6.3 Polymer chemistry

The mention of synthetic fibres brings us to what must be one of the most striking developments in chemical technology over the past thirty years, the field of man-made bulk materials. These include paints, insulation plastics and foams, rubbers, plastics, perspex and many new building materials, as well as the synthetic fibres such as nylon, 'terylene', etc. All are composed of very large molecules, each made up of identical sub-units in much the same way as cellulose is made up of linked glucose units. Such materials are known as polymers. Those manufactured have very special properties and their design and synthesis has been a direct consequence of an understanding of the dependence of chemical and physical properties on molecular structure. Some synthetic fibres are far superior in many respect to any natural fibre. The synthesis, structure and properties of polymers will be discussed at greater length in Unit 13.

10.6.4 Pharmaceutical chemistry

The development of drugs in the last century has made an enormous impact on humanity. The lengthening of the average human life span has implications for the future, of which most of us are aware (and to which we shall return briefly later in this section and in Unit 20). It is therefore of considerable interest to examine a few examples of the contribution the molecular scientist has made to this field. Because biochemical mechanisms associated with disease are still largely unknown, most of the development has been along semi-empirical lines. Typically, a naturally occurring or man-made substance has been found, through long established application or by chance discovery, to have a beneficial effect on people afflicted with certain ailments. Once the structure of the compound is known, the chemist has set to work and synthesized a huge number of structural variants until one is found that is more effective and cheaper to produce than the first material and, furthermore, has no undesirable side effects. Such was the case, for example, in the synthesis of the widely used anaesthetic, Novocain. Early explorers found that the Indians of Peru munched coca leaves in order to numb sensations of pain. Subsequently, the active principle of the leaf, cocaine, was isolated and its structure determined. Although it could then be synthesized in the laboratory, manufacturing it in this way would have been far more expensive than growing coca leaves and extracting it chemically.

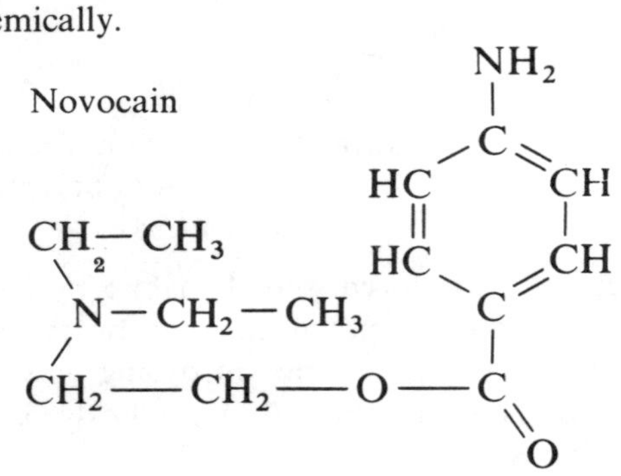

However, it seemed that there was no reason why compounds with structures related to cocaine and with more powerful anaesthetic action should not exist. Moreover, cocaine itself has a serious disadvantage when it is used in man.

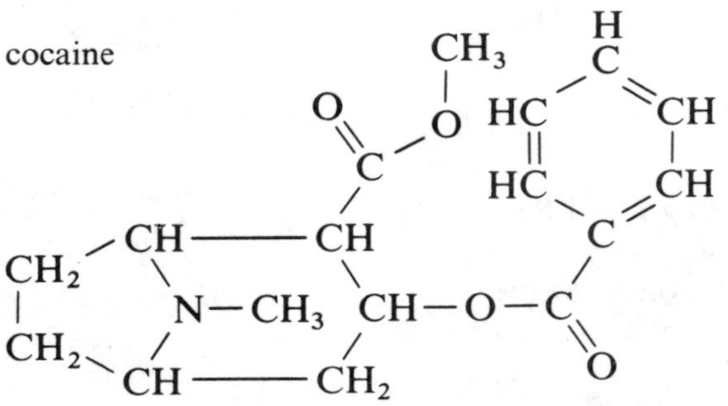

Do you know what it is?

Cocaine is addictive. The results of the search by the synthetic chemist for a suitable alternative led to the discovery of Novocain which, though not as powerful as more recently developed derivatives, has low toxicity, is not addictive and is easy to handle.

Novocaine falls into one of two classes of drugs, the pain-killing class. The other, the chemotherapeutic* class, is used in the control of parasitic infections and includes anti-biotics such as the penicillins, obtained largely from natural sources, and the sulpha drugs (typically, the sulphonamides shown below) which are synthesized from coal tar and petroleum products.

Both types of antibiotic were discovered initially by chance† observation, and in both cases the effects of structural modifications have been explored. The result in the case of penicillin has been the production of a derivative which can be administered orally (thus, greatly expanding the ease of penicillin therapy) and, in the case of the sulpha drugs, a versatile range of agents which can be used for a wide variety of infections.

* *Pertaining to the treatment of disease by chemical means. Chemotherapeutic drugs are used for purposes other than the treatment of infectious disease. In fact, in terms of production their use for clinical conditions such as diabetes, nervous disorders, depression etc. far outweighs that for diseases caused by harmful bacteria.*
† *Not completely by chance, however, in the case of penicillin. The observation was made during a research programme whose aim was to find such an anti-bacterial agent. The 'chance' refers to the contact of a particular mould with the bacteria being used in the investigation.*

Penicillin G

sulphanilamide

sulphapyridine

sulphathiazole

10.6.5 Agricultural chemistry

Attempts to control disease and other factors which affect plant growth have led to a large modern pesticide industry. Pesticides fall into three classes; fungicides, herbicides and insecticides.

In the mid-nineteenth century, over one million people died in Ireland as a direct result of a fungus which appeared as dark blotches on the leaves of potato plants and all but wiped out the main source of food in that country at the time. Fungicides now account for a very large portion of agricultural chemicals.

Food supply from plants would be diminished if unwanted plant growth, which competes with crops for fertilizer and water, were not removed. The manufacture of herbicides has given man considerable control over weeds.

Finally, the destruction of plants by insects has a calamitous effect in certain areas of the world. Locusts come easier to mind than most, though there are many other pests which are almost as destructive. The discovery of D.D.T. was a landmark in insect control and led not only to improvements in crops, but to control of certain insect-borne diseases, notably malaria.

D.D.T.

The approach of the chemist to the synthesis of these materials was similar to that followed in the pharmaceutical industry. Once an effective substance was discovered, a large number of structurally similar compounds were made and tested. From such an approach it was found, for instance, that compounds containing the thiocarbamate group were effective to varying degrees against fungus, as also were organochlorine compounds. (Carbon compounds which contain chlorine.) Organochlorine compounds were also found to be effective herbicides and insecticides.

a thiocarbamate

dieldrin, an organo-chlorine insecticide

malathion, an organo-phosphorus insecticide

Carbamates and organophosphorus compounds (whose properties were discovered through the study of the nerve gases) were also found to be powerful herbicides and insecticides respectively.

Pesticides, by their very nature, are toxic to life. A good agent must leave the host plant unaffected and its lifetime in the natural environment must be sufficient to produce the desired effect on the pest but insufficient to produce adverse effects on other forms of life. In this sense, the molecular design is important because man now knows enough chemistry to be able to incorporate into the molecular structure of a compound features which will limit its lifetime to a known extent. However, much knowledge of the relationship between living organisms and their surroundings (ecology) is needed in order to develop agents with the required selectivity. The implications of agricultural chemicals will be discussed further in Unit 20.

10.6.6 Detergents

Detergents may be described as substances capable of dispersing in water various water-insoluble substances such as grease and dirt. They have been known for over two thousand years and, until quite recently, were obtained from animal and vegetable fats. Treatment with alkali cleaves a fat molecule into glycerol and the sodium salts of long-chain fatty acids (soaps such as sodium stearate).

$$CH_3CH_2CH_2CH_2CH_2CH_2CH_2CH_2$$
$$CH_2$$
$$CH_2CH_2CH_2CH_2CH_2CH_2CH_2CH_2$$
$$C$$
$$O \diagdown O^- Na^+$$

sodium stereate,
a soap

$$CH_2-OH$$
$$CH-OH$$
$$CH_2-OH$$

glycerol

The principles behind the detergent action of soaps have been known for some time and follow from the rule of thumb 'like dissolves like'. As you can see, soap molecules consist of a long (non-polar) hydrocarbon chain terminated by a (very polar) ionic group. The potential for solubility in oil *and* water is incorporated into the same molecule. When an aqueous solution of a soap makes contact with an oil or grease film, the hydrocarbon end tends to lodge itself in the latter leaving the polar group in the water. Electrostatic repulsion between these polar groups leads to a breaking up of the grease layer into globules which effects dispersion and hence cleansing of the surface to which the grease was originally bound.

Although the sodium soaps are soluble in water, those of calcium, magnesium and other ions found in hard water are not, and come out of solution as scum. The use of soap in hard-water areas is thus extremely wasteful. This difficulty, together with the shortage of animal and vegetable fat during the Second World War, stimulated research which ultimately led to the synthetic detergents or 'syndets'. They were designed according to the principles discussed above, but they have additional virtue that they do not form insoluble salts with calcium, magnesium or ferric ions. For this reason, they may be used with essentially the same efficiency in hard as in soft water. Many of those marketed today have similar hydrocarbon chains to soaps, but have what are known as sulphate ester end groups instead of carboxylic end groups.

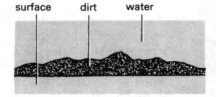

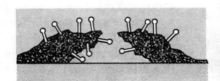

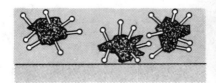

Figure 24   Detergent action.

$$CH_3.(CH_2)_{16}.CH_2.O.SO_3^- Na^+$$

a sulphate ester syndet

The hydrocarbon fragment is obtained in the form of a hydroxy-compound from fats cleaved with hydrogen. Treatment with sulphuric acid and sodium hydroxide then produces the detergent.

$$\begin{array}{c} CH_2.O.CO.R. \\ | \\ CH\ .O.CO.R. \\ | \\ CH_2.O.CO.R. \end{array} \xrightarrow{H_2} \begin{array}{c} CH_2.OH \\ | \\ CH\ .OH. \\ | \\ CH_2.OH. \end{array} +\ 3R.CH_2.OH \xrightarrow[(2)\ NaOH]{(1)\ H_2SO_4} R.CH_2O.SO_3^- \ Na^+$$

Other types of syndet incorporate a benzene ring and the sulphonate group.

When syndets became very widely used, cheaper methods for synthesis were developed, of which the most useful involves the formation of the hydrocarbon chain by partial polymerization of the petroleum products. Detergents made in this way prove to be very effective cleansing agents. The original compounds developed, however, persisted in sewage effluent and water reservoirs to the extent that at one point large stretches of the River Trent, for example, became sudsy! In some areas in the United States even the drinking water supply in the domestic household was affected.

The fatty acid hydrocarbon chain is linear and capable of degradation by bacteria present in sewage plants and soil. Such bacteria appear to be 'baffled' by the branched hydrocarbon chains which resulted from the early petroleum product polymerization processes.

$$-CH_2-CH_2-CH_2-CH_2-CH_2-$$

biodegradable

$$CH_3-\underset{\underset{CH_3}{|}}{CH}-CH_2-\underset{\underset{CH_3}{|}}{CH}-CH_2-$$

not biodegradable

Nowadays, processes are available for the economic manufacture of straight chain material from petroleum products, and biodegradable detergents have largely superseded those that are more obviously socially undesirable.

$$\begin{array}{c} CH_3 \\ CH_2 \\ CH_2 \\ CH_2 \\ CH_2 \\ CH_2 \\ CH_2 \\ CH_2 \\ CH_2 \\ CH_2 \\ CH_2 \\ CH_2 \\ | \\ C \\ \end{array}$$

a benzene-
sulphonate syndet

This last sentence gives you a clue as to what ought to go into the box marked X in Figure 23 (p. 47). Can you think what it should be?

The query 'socially acceptable?'

10.6.7 New problems

The topics we have been discussing in this section demonstrate in a rather superficial way some of the potential of man's increasing knowledge of structure-property relationships and chemical reactions at the molecular level. The very advances that can be made as a result, however, produce new problems. The balance of chemical and physical processes in life on our planet (the biosphere) is a very delicate one. On the one hand, death due to bacterial infection has been brought under control and the life span of the average human being greatly lengthened. On the other, this control of death has in part been responsible for the current phenomenal population explosion, and has focused attention on such problems as food output and birth control. Fortunately, these problems are also proving to be capable of chemical control. Thus, at one and the same time, organic chemistry holds out the promise of postponing death, regulating births and improving the material quality of life.

Again, the ever increasing demand for higher food production has brought about the widespread use of pesticides. Indeed, without their use, the total agricultural production of the world would be too low to sustain even the present population. Unfortunately, in certain areas their application is adversely affecting the environment and this, together with problems of pollution caused by the vast industrial growth of the developed countries, will make increasing demands on the chemist and other scientists in the future. Indeed it may be said that the great challenge facing the modern scientist and technologist is that of sustaining an improvement in man's environment within the context of a biospheric and geophysical equilibrium. As far as the chemist is concerned, this will ultimately require a greater understanding of chemical and physical behaviour, and the way it is linked to molecular structure.

(The social implications of scientific and technological progress will be discussed at length in Units 33 and 34.)

> **Summarize what you consider to be the important points from section 10.6 and compare your summary with that given below.**

10.6.8 Summary of section 10.6

1. See Figure 23 for a summary of factors involved in chemical manufacture.

2. Dyes must possess structural features which
    - (i) make them coloured (section 10.5.4),
    - (ii) combined with those of the fabric and method of application, promote strong adhesion,
    - (iii) are stable under the conditions the dyed object is likely to encounter.

3. Lack of knowledge of the biochemical mechanisms associated with disease has lead to a semi-empirical approach to drug manufacture. Applications of theoretical knowledge of structure/property relationships and chemical reactions are usually involved after initial empirical observations of the effects of a particular type of chemical compound.

4. Similar remarks apply to the manufacture of agricultural chemicals particularly pesticides.

5. A detergent molecule incorporates both a long hydrocarbon chain and a polar group. This allows it to dissolve in both oil/grease and water and facilitates cleansing by dispersion. The design of modern synthetic detergents is based on this structure/property relationship and also takes into account the observation that only straight chain hydrocarbon fragments are biodegradeable.

Recommended Parallel Reading

Nuffield Advanced Science: Chemistry, Book I, Background reading sections of Topic 9, pp. 290, 293, 299–303, 305–309.

K. Mellanby, *Pesticides and pollution*, Fontana, 1969. (A background reading reference for Unit 35).

SAQS relevant to section 10.6 are numbers 23 and 24.

10.7 Summary

10.7.1 Covalency in carbon compounds

It was shown in Unit 8 that the position of an element in the Periodic Table determines whether it will tend to form ionic or covalent compounds. The structures of covalent molecules are examined in this Unit, through a study of the compounds of the element, carbon. This approach is taken because:

(i) the behaviour of the compounds of carbon is typical of those of covalent compounds;

(ii) the known compounds of carbon far outnumber the covalent compounds of all other elements put together;

(iii) carbon occupies a uniquely important role in nature as being the basic element of all living systems.

A carbon atom can attain the stable rare-gas electronic configuration of neon by sharing its four valence electrons with other atoms in such a way that it becomes surrounded by four electron pairs in its valence shell. It does this very effectively with hydrogen atoms and other carbon atoms, forming strong bonds with both elements to give the hydrocarbons. The diversity in the structures so produced is already apparent because the only restriction is that each carbon atom must participate in four bonding electron pairs and each hydrogen atom one. Thus straight chains, branched chains and rings are possible. For a given number of carbon and hydrogen atoms it is possible to have many different structures. Further variation in structure arises from the sharing of not only two but of four and six electrons between carbon atoms (multiple bonds).

10.7.2 The uniqueness of carbon

Carbon is compared with its nearest neighbour in the Periodic Table, silicon. The property which dominates silicon chemistry and the role it plays in nature is the strength of the Si—O bond. Unlike the C—C bond, the Si—Si bond is rather weak. Moreover carbon forms strong bonds with many elements in addition to itself and hydrogen.

Thus the diversity in structure expected of carbon compounds becomes further increased when one includes the formation of stable compounds consisting of carbon chains or rings to which may be attached hydrogen, oxygen, nitrogen, chlorine and so on. This variety in molecular structure inevitably leads to variety in chemical and physical behaviour, and was no doubt responsible for the emergence on this planet millions of years ago of a group of molecular species which were able to mutually self replicate and ultimately form the basis of life as we know it (chemical evolution). Prior to 1828, living or dead organisms were the only source of carbon compounds (organic compounds). In that year, Wohler synthesized urea from inorganic substances. At the present time, organic compounds synthesized in the laboratory far outnumber those isolated from natural sources.

10.7.3   The shapes of organic molecules

The shapes of covalent molecules can be predicted on the basis of minimal repulsion between bonding and between non-bonding electron pairs in the valence shells of the atoms concerned. Thus the tetrahedral configuration of saturated carbon, which has four such electron pairs, is established and bond angles in most simple covalent molecules broadly rationalized. We are now in a position to visualize the three dimensional structures of simple organic molecules and apply the principles involved to larger more complex systems. Factors which determine not only the shape but the flexibility of a molecule are established. The vulnerability of electrons in multiple bonds (polarizability) to external reagents accounts for the relative chemical instability of unsaturated compounds.

So far the principles governing molecular shape have been exemplified with simple organic molecules of random origin. Further illustration is now made using well-known naturally occurring substances chosen in part for their relevance later in the course. Thus the structure of glucose and the primary structures of proteins are examined. The latter leads into a subtle but vitally important aspect of molecular shape—chirality (handed-ness or asymmetry)—and the way it is monitored through optical activity.

10.7.4   The influence of structure on physical and chemical properties

The explosive growth of knowledge in this field has not only enabled man to understand the behaviour of much of the natural material world about him, but has facilitated the synthesis of new materials designed to fulfill specific roles in his attempts to improve his environment.

The physical and chemical behaviour of every organic compound are dictated by its geometric and electronic structure. We have just discussed the former and have met the latter (polarization and polarizability) briefly in earlier sections. Some simple physical properties (m.p., b.p., solubility and colour) of a representative sample of organic compounds are examined. Chemical reactivity is identified with the tendency for a molecule to undergo changes in bonding, with or without the assistance of another molecular species. Here the dominating influence is the elec-tronic aspect of structure, and the ease with which bonds are formed or broken is related to polarization and polarizability.

The role of geometric shape (stereochemistry) in determining chemical reactivity is examined briefly and exemplified by some simple cases, and by enzyme catalysis.

10.7.5   The influence of structure on properties — examples from chemical technology

Although dyes, polymers, agricultural chemistry and the petroleum industry is mentioned, attention is focused on pharmaceutical chemistry and detergents. In particular, the role of molecular design in modern chemical technology is exemplified by the synthesis of the non-addictive anaesthetic Novococaine, and the modelling of biodegradable synthetic detergents on the structures of those obtained from animal and vegetable fats (soaps).

Finally the role of the molecular scientist in solving the problems that modern technology has created is briefly discussed.

Book List

Recommended parallel reading referred to in text:

M. J. Sienko and R. A. Plane, *Chemistry: Principles and properties*, McGraw-Hill, 1966.

D. H. Andrews and R. J. Kokes, *Fundamental Chemistry*, Wiley, 1965.

S. Dunstan, *Principles of Chemistry*, Van Nostrand, 1968.

Nuffield Advanced Science: Chemistry, Book 1, Penguin Education, 1970.

Chemical Bond Approach Project, *Chemical Systems*, McGraw-Hill, 1964.

Appendix 1

Ball and Spring Molecular Models

10.A1.1 Introduction

Although one can represent the (three-dimensional) shapes of molecular structures by (two-dimensional) drawings, they are best understood by the examination of molecular models. These can be built from components that have been designed to give shapes whose geometries and behaviour closely resemble those of real molecules as deduced from experimental observations. In fact, molecular models are indispensable for acquiring a sense of the true shapes of organic molecules, their flexibility or rigidity and the way in which the shapes can be changed by 'rotation' of the atoms around the bonds. There are many types of models available; each have their own advantages and disadvantages and all have limitations. We have chosen the ball and spring variety for this Unit.

10.A1.2 Description

In your Home Experiment Kit you will find a plastic box containing the following balls and springs:

Number	Element	Colour	Holes	Hole configuration
12	hydrogen	white	1	—
6	carbon	black	4	tetrahedral
6	carbon	black	14	universal
6	oxygen	red	2	dihedral as O in H_2O
2	nitrogen	blue	4	tetrahedral
2	halogen	green	6	octahedral
21	38 mm spring bonds (C—C, C—O, C—N, C—Cl)			
12	25 mm spring bonds (C—H, O—H, N—H)			

The different coloured balls represent atoms of different elements and the internationally agreed colour/element convention has been adopted in the above table. The colours, however, are unimportant compared with the way in which the holes are drilled. These determine the 'bond angles' within the model, and their configuration is chosen to match the observed bond angles in real molecules. Slight angle distortions from those fixed by the holes can be accommodated by using stiff springs for bonds instead of rigid rods. The different lengths of the springs represent the major differences observed in interatomic distance, that between the X—H, C—H type and the C—X, C—C type, where X is an element of the first row of the Periodic Table. The springs you have with your set correspond to a scale of 35 mm = 1 ångström ($= 10^{-10}$ metre).

.A1.3   Limitations

(i) The relatively large size (necessary for their construction) of the balls representing the nuclei is not really representative of molecular geometry. Although nearly all the mass of a molecule is in the atomic nuclei, the space it occupies is small compared with the interatomic distances. If we use the full stop after the last sentence to represent the nucleus of a carbon atom, then a carbon carbon bond would be about fifteen metres long!

(ii) You should be able to rotate balls connected by a single spring, and any stiffness you encounter when you try to do this is a feature of the model and not the molecular fragment it is supposed to represent. Resistance to rotation becomes a serious handicap in the construction of rings involving five or more atoms. (The shapes these rings adopt in real molecules are generally those which cause minimum distortion of normal bond angles, and minimum across-space repulsion between bonding electrons). You should try to minimize this handicap by rotating the bonds in their holes as you construct the models, though do not do it too vigorously, or they may fall apart too easily once made.

(iii) Organic molecules strongly resist forces which alter their bond angles from normal values. Your models will resist this too, but not to the same extent. Consequently, they can only give you a very rough idea of stability when it comes to making very strained molecules. On the one hand, you may find it very easy to make a model of a molecule which would be too unstable to exist, or on the other, very difficult to construct one of an extremely stable molecule.

Take care when dismantling models. Springs can be extricated from the balls by pinching them near the hole with the thumb and bent forefinger, as shown. *On no account use pliers for this operation, as you may distort the springs.*

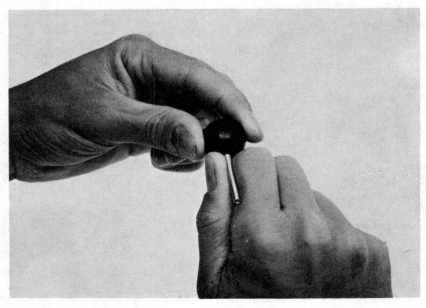

Figure 25   How to take the spring out.

Appendix 2

Plane-Polarized Light

Unit 2 described light in terms of a wave in which the magnitude of the electric and magnetic fields perpendicular to the direction of propagation vary in a regular fashion.

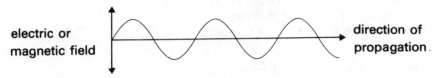

electric or magnetic field ——→ direction of propagation

Figure 26   A wave.

Thus, activity in two dimensions is well-defined, but we live in a three dimensional world! The most we can say is that the changes in the electric and magnetic fields take place in a *plane* at right angles to the direction of propagation and, in fact, in ordinary light the direction within this plane is random (Fig. 27a). Certain substances, however, known as polarizing materials (e.g. 'polaroid') can reduce such light to a form in which the fields change only in one plane and the light is then said to be *plane-polarized* (Fig. 27b). If now another piece of the material is placed in the emerging beam, and is oriented in precisely the same way as the first, then the light it transmits is unchanged: that is, it remains plane-polarized. However, if the second piece is oriented at right-angles to the first, then no light emerges from the end of the system (Fig. 27c). You can observe this yourself with the two pieces of polaroid film in the Home Experiment Kit. View daylight or artificial light through both pieces and then rotate one relative to the other. (Sunlight reflected off many surfaces is partially plane-polarized and this accounts for the use of polarizing film in sun-glasses).

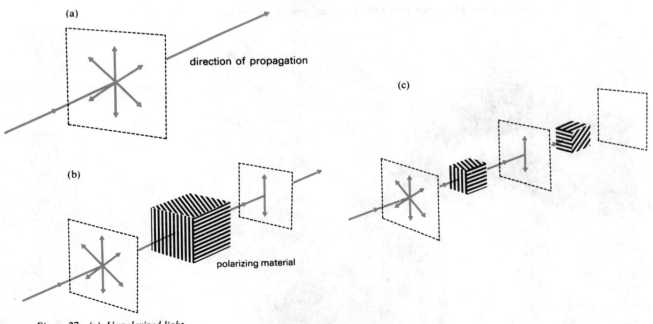

(a)

direction of propagation

(c)

(b)

polarizing material

Figure 27   (a) Unpolarized light
(b) Plane-polarized light
(c) Extinction of light by polarizers set at right angles.

The Orbital Approach to the Geometry of Organic Molecules

Although electron repulsion theory is adequate for our purposes for explaining the shapes of organic molecules in the Foundations Course, it is not the approach taken by most authors of modern chemistry texts. Accordingly it would be an advantage to those of you who anticipate taking higher level courses in chemistry to become aware of the somewhat more complex approach based on the concept of atomic orbitals. This is described in some of the texts we have recommended for additional reading:

M. J. Sienko and R. A. Plane, ibid., sections 2.3, 3.8, and, if you have time, 3.9. Chemical Bond Approach Project, ibid, sections 10.17 to 10.23 incl. S. Dunstan, ibid., pp. 14, 67–69. D. H. Andrews and R. J. Kokes, ibid., pp. 83–101.

Distortion Factors of Electron Repulsion Theory

Examination of the series 1, 2, 3 of Table 2 (p. 25) shows the hydrogen atoms being pushed closer together as unshared electron pairs are introduced. Thus *non-bonding electrons appear to occupy more space than bonding electrons.*

Can you offer an explanation for this?

Unshared electrons are, by definition, likely to be held closer to the nucleus than shared electrons. Consequently, they take up more room on the surface of the atom.

Compounds 7, 8 and 9 in Table 2 contain double bonds and the carbon atoms in each case are bound directly to three atoms, not four. In this sense, they are analogous to elements of Group III in the Periodic Table, e.g. boron. Boron trifluoride has three identical electron pairs in its valence shell and its molecular shape is trigonal, the F—B—F bond angle having the trigonal value of 120 degrees. Now we would expect *the 4 or 6 electrons of multiple bonds to occupy more space than the 2 electrons of single bonds.* This is supported by evidence from Table 2. All the X—C—X bond angles of compounds 7, 8 and 9 are less than 120°, showing that the double bond is taking up extra space, pushing the C—X bonds closer together.

There is one more important factor which must be taken into account before all the variations in Table 2 can be explained. Consider the pairs 2, 4 : 3, 5 and 7, 8.

The table shows that repulsion between shared electron pairs decreases as they are drawn away from the central atom. Fluorine is more electronegative than hydrogen and so the bonding electrons of nitrogen trifluoride are found at greater average distance from the nitrogen atom than those of ammonia. The repulsion between the bonding electrons is less and hence so is the bond angle. Sulphur is less electronegative than oxygen so, once again, the bonding electrons in hydrogen sulphide repel one another less than those in water. (However, we must be careful here as we are talking about two different valency shells, sulphur lying in the second Period of the Periodic Table.)

Comparison of the X—C—X bond angles in 7 and 8 shows the same trend as is apparent with 2 and 4.

(a)

(b)

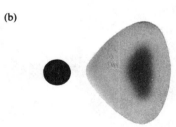

Figure 28   Relative distribution of bonding (a) and non-bonding (b) electron pairs.

What about 9, though? Is this anomalous or consistent?

Oxygen is more electronegative than carbon so we might expect electron repulsion between the C=O and C—F bonds in 9 to be less than that between the C=C and C—F bonds in 8. This leads to the conclusion that the O—C—F bond angle should be smaller than the C—C—F bond angle whereas the opposite appears to be the case. The situation is anomalous and the explanation not clear. We must take note, however, that the difference here is quite small compared with other differences in the Table.

The Structure of Benzene

The unsaturated compound benzene has a molecular formula C_6H_6. Experimental evidence shows that the six carbon atoms form a flat six-membered ring.

See if you can draw a two-dimensional structure for benzene consistent with reasonable bond angles.

Your answer should have been one of the equivalent structures (a) and (b) in Figure 29 (p. 64). If it included two adjacent double-bonds such as (c), construct the grouping $C{=}C{=}C$ with your models and note the natural shape of this system. Now make (a) with your models.

We know from experiment that all six bonds in benzene are equivalent having a length intermediate between those of $C{=}C$ and $C{-}C$.

Is your model consistent with this?

It shouldn't be!

Furthermore, the chemical behaviour of benzene is radically different from open-chain unsaturated systems. In particular, its chemical stability is considerably greater than would be predicted on the basis of structures (a) and (b) (Fig. 29). In fact, the picture we have created of single and double bonds breaks down in the case of benzene.

Kekulé, who as long ago as 1865 formulated structures (a) and (b) for benzene, got over the problem of stability by suggesting that the bonds in (a) and (b) alternated position so rapidly that each carbon carbon bond was neither single nor double but something intermediate. He once related how the idea occurred to him.

> I was sitting, writing at my text-book; but the work did not progress; my thoughts were elsewhere. I turned my chair to the fire and dozed. Again the atoms were gambolling before my eyes. This time the smaller groups kept modestly in the background. My mental eye, rendered more acute by repeated visions of the kind, could now distinguish larger structures, of manifold conformation: long rows, sometimes more closely fitted together; all twining and twisting in snake-like motion. But look! What was that? One of the snakes had seized hold of its own tail, and the form whirled mockingly before my eyes. As if by a flash of lightning I awoke; and this time also I spent the rest of the night in working out the consequences of the hypothesis.
>
> Let us learn to dream, gentlemen, then perhaps we shall find the truth . . . but let us beware of publishing our dreams before they have been put to the proof by the waking understanding.

The study of benzene and other compounds which have similar properties has resulted in a generally accepted rule that where two equivalent alternative bond structures can be drawn the actual structure is neither of these. In such molecules, the available electrons are distributed evenly between the atoms involved. The only geometric resemblance of the model you have made, to the benzene molecule is the fact that all the atoms lie in one plane. One popular picture of benzene is that in which each carbon atom is bonded to the next by two electrons forming a single covalent bond, the remaining six electrons not existing as three pairs, but distributed as charge clouds above and below all the atoms of the ring.

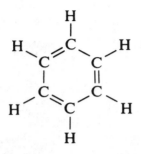 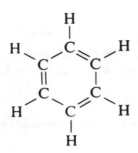 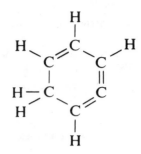

(a) reasonable
     bond
     angles

(b) reasonable
     bond angles

(c) unreasonable
     bond
     angles

Figure 29   Possible structures of C_6H_6.

These electron clouds are mobile and the electrons of which they are composed are known as *delocalized electrons*. The tetrahedrally drilled balls are not capable of showing this structure, but we can get closer to it using the universally drilled balls. Construct benzene as follows. Insert one 25 mm spring into hole 1 in each of six universally drilled black balls.

delocalized electrons

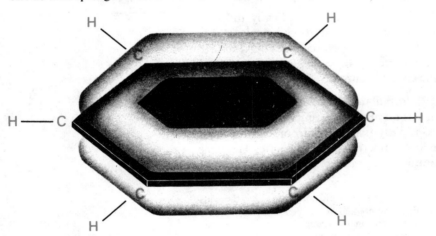

Figure 30   A possible electronic structure for benzene.

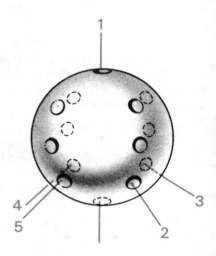

Figure 31   A universally drilled ball.

Connect the six balls together in one ring by pairs of 38 mm springs using holes 2, 3 and 4, 5. You will find this requires some dexterity! The model you finish with should look like the one in Figure 32 (plate facing p. 96).

Compare the first model you made with the one just completed. In the former, one spring represents a pair of bonding electrons. We have had to abandon this idea in trying to get close to the true structure of benzene. However, we now have a structure where all the carbon carbon bonds are equivalent.

Is the carbon carbon bond distance consistent with that observed experimentally?

In much the same way that our models are somewhat inadequate for benzene, so is our convention for writing down organic structures where we represent one pair of electrons by a single line. We know that structures

You remember we said that the carbon carbon bond length in benzene is intermediate between that of a normal single bond and a normal double bond. Your model should be consistent with this.

(a) and (b) are wrong, because they each imply two sets of different bonds. One way around this is to represent the benzene ring as shown below. However, structures such as (a) and (b) are sometimes more convenient, particularly when dealing with chemical reactions as opposed to structure, and they are frequently used in chemical literature.

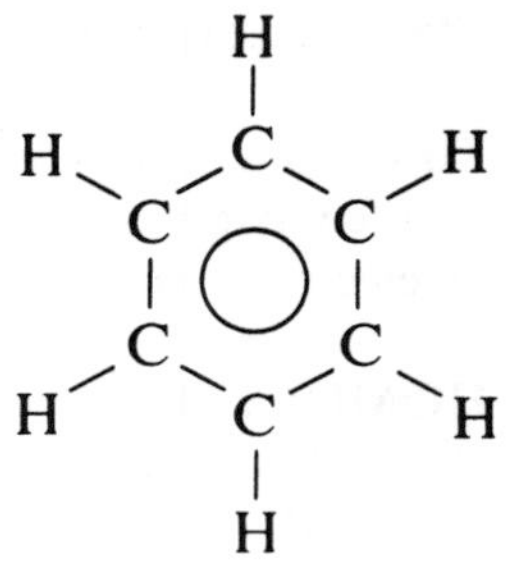

An explanation as to why the electrons of benzene behave in this manner is beyond the scope of the course. Nevertheless, the unique stability of a set of six delocalized electrons within a ring can be regarded as a molecular analogy to the unique stability of the closed shell electronic structures of the rare gas elements. This concept in organic chemistry is termed *aromaticity* and compounds which contain this feature are known as *aromatic* compounds. (It appears that certain benzene derivatives were obtained originally from oils with pleasant aromas, and as they were of unknown constitution they were classed together as aromatic compounds.)

aromaticity
aromatic compounds

Delocalized electrons are not confined to aromatic species. Many covalent groups have structures which are not adequately represented by the dash/bonding pair notation. One such group is the carboxylate anion grouping $-CO_2^-$, formed typically by the action of an inorganic base with acetic acid.

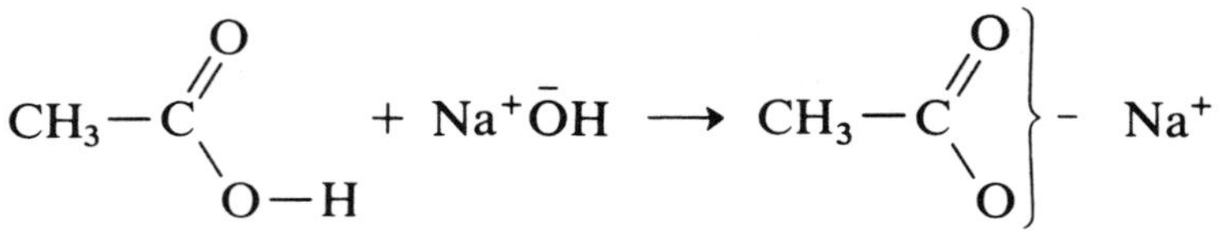

XXXII

XXXIV

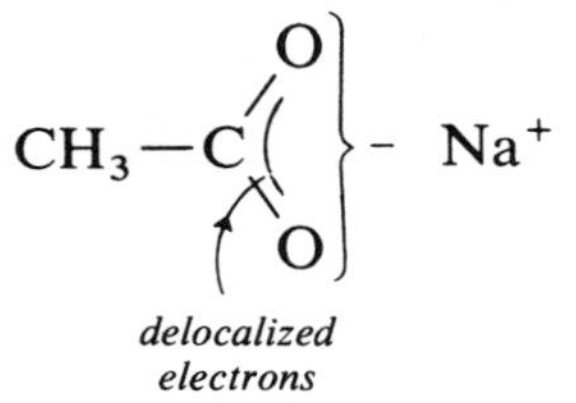

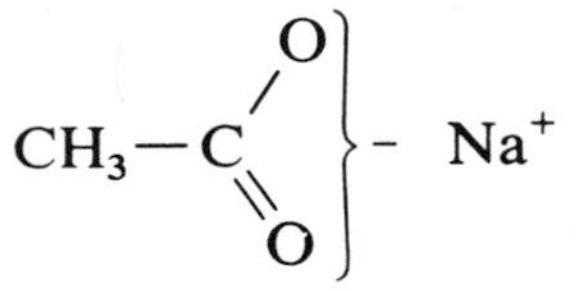

delocalized
electrons

XXXIII

An equivalent structure for XXXII is XXXIII and so the true structure is best represented by XXXIV.

Write down a reasonable structural representation of sodium benzoate $C_6H_5CO_2^-$ Na^+, in which one hydrogen atom of benzene has been replaced by $-CO_2^-$ Na^+.

See Answer 7, p. 95.

Recommended parallel reading

Nuffield Advanced Science: Chemistry, Book 1, Topic 8, pp. 252–257.
D. H. Andrews and R. J. Kokes, ibid, pp. 99, 100.

Electronic Structure and Chemical Reactivity

This is an enormous subject and we can only give it superficial treatment in a few pages. Let us start, however, by recognizing two ways by which an electron pair bond can be formed or broken.

HOMOLYTIC
CLEAVAGE or $\quad\quad\quad$ $A \cdot \vdots \cdot B \longrightarrow A\cdot + B\cdot \longrightarrow$ etc.
HOMOLYSIS

HETEROLYTIC $\quad\quad$ $\begin{cases} A \vdots \cdot\cdot B \longrightarrow A^+ + B^- \longrightarrow \text{etc.} \\ A \cdot\cdot \vdots B \longrightarrow A^- + B^+ \longrightarrow \text{etc.} \end{cases}$
CLEAVAGE
or HETEROLYSIS

The reaction need not result in atoms with unpaired electrons or ions, in fact most chemical reactions do not stop at this stage, but consist of a series of bond forming and bond breaking processes. Furthermore, bond making can take place at the same time as, or even after, bond breaking:

$$C + AB \longrightarrow CA + B.$$

The nature of C, the A—B bond and the reaction conditions, then, determine whether the overall process will occur homolytically or heterolytically.

Consider the transformations listed below and suggest whether changes in the bonding take place heterolytically or homolytically.

(1)–(3) homolytic
(4)–(6) heterolytic

$Cl-Cl \xrightarrow[\text{gas phase}]{\text{sunlight}} Cl\cdot + Cl\cdot$ ....(1)

$Cl\cdot + CH_4 \longrightarrow CH_3^\cdot + HCl$ ....(2)

$CH_3^\cdot + Cl-Cl \longrightarrow CH_3Cl + Cl\cdot$.................(3)

$CH_3Cl + Na\overset{+}{O}\overset{-}{H} \xrightarrow{\text{water}} CH_3OH + NaCl$ ....(4)

$H-Br + CH_2{=}CH_2 \xrightarrow[\substack{\text{water} \\ \text{room temperature}}]{CH_3CO_2H} \overset{\displaystyle H}{\underset{\displaystyle CH_2-CH_2-Br}{|}}$ ....(5)

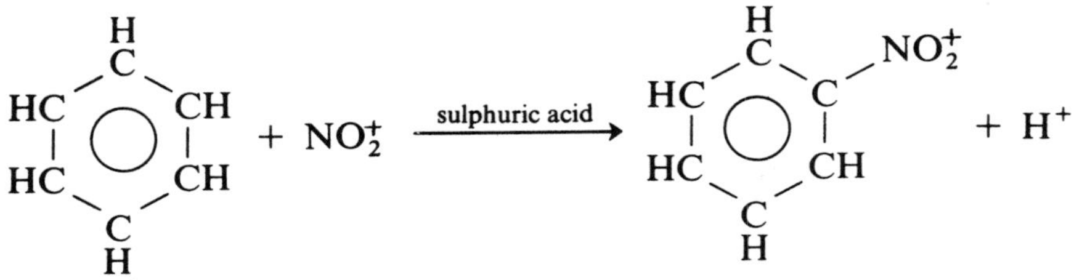

In our general reaction C + AB etc., it appears that, as far as C is concerned, we expect charged or polar reagents to participate in heterolytic reactions, whereas reagents with unpaired electrons (free radical reagents)

should participate in homolytic reactions. As far as the bond between A and B is concerned, polarization tends to induce heterolysis, whereas a non-polar bond tends to break homolytically. The conditions are also important. A polar solvent (water or acetic acid, CH_3CO_2H, for example) tends to induce heterolysis. This is discouraged in a non-polar solvent or in the absence of solvent, when homolytic reactions may occur.

Chemical reactions involve the reorganization of electrons within a molecule or molecular fragment. Clearly, the more loosely bound or more mobile the electrons are, that is, the more polarizable they are, the greater the ease with which such reorganization can take place. The carbon carbon bond in ethane $CH_3—CH_3$, is very strong. Its electrons have low polarizability and only under very drastic conditions can it be ruptured. The electrons of the carbon carbon bond of ethylene, $CH_2=CH_2$, however, are quite polarizable and, as you saw a moment ago in reaction 5 (p. 66), one of the two electron pair bonds can be cleaved under quite mild conditions. All multiple bonds, in fact, are similarly polarizable. One class of compounds, however, is less chemically reactive than one would expect from the polarizability of its electrons.

Can you draw a typical structure of a compound in this class?

Compounds containing benzene rings do not normally undergo the same kind of reaction as say, ethylene or formaldehyde, $H_2C=O$, although they can be quite reactive in other respects. They tend to interconvert in such a way as to preserve the stable multiple-bonded arrangement, the aromatic ring.

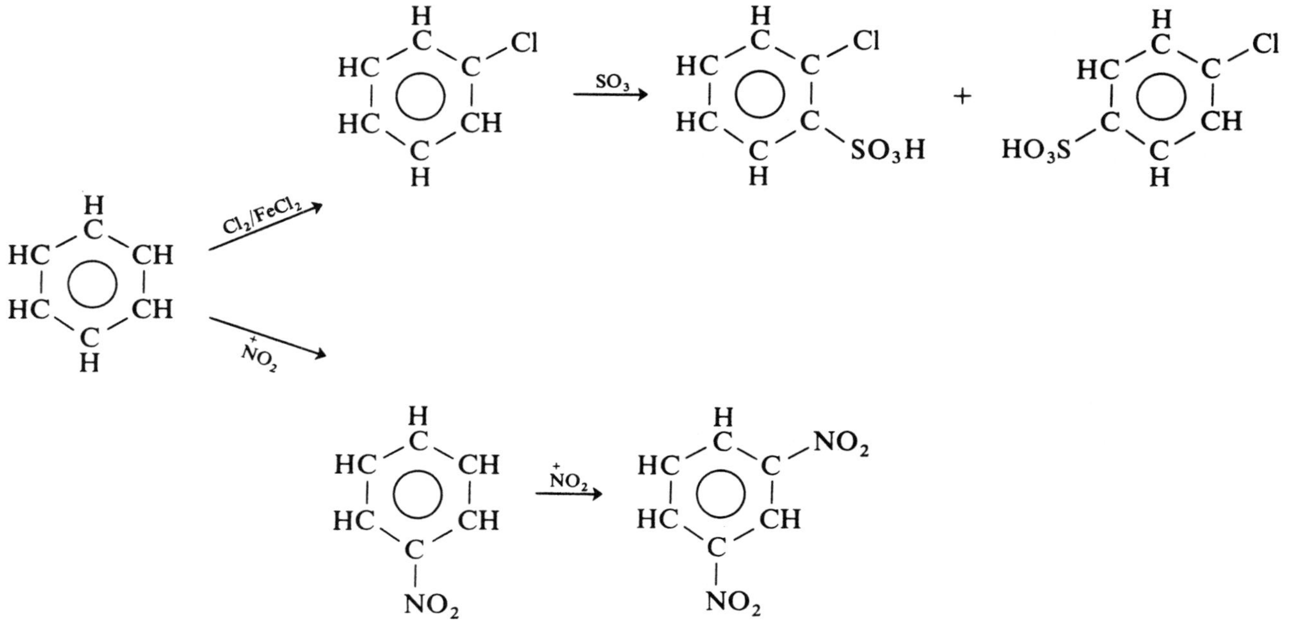

We said earlier that non-bonding electron pairs are also polarizable. Indeed, they constitute the source of reactivity in many organic compounds by forming bonds to positively charged or electron deficient reagents.

$$CH_3-\ddot{O}-\bigcirc \xrightarrow[H^+]{H_2O} CH_3.OH \;+\; :\ddot{O}-\bigcirc \quad \ldots(7)$$

XXXV

via

$$\left\{ \;CH_3-\overset{+}{\underset{\underset{H\diagdown\ddot{O}\diagup H}{\;}}{O}}-H\phi \;\;\longrightarrow\;\; \underset{H}{\overset{CH_3}{\underset{\diagup}{O^+}}}\diagdown H \;+\; :\ddot{O}-\bigcirc \right.$$

$$\left. \xrightarrow{-H^+} CH_3.OH \right\}$$

and

$$\underset{CH_3}{\overset{CH_3}{\diagdown}}HN: \;+\; CH_3-\overset{C}{\underset{\;}{\overset{\|}{C}}}-Cl \;\longrightarrow\; CH_3-\overset{O}{\underset{\;}{\overset{\|}{C}}}-\ddot{N}\overset{CH_3}{\underset{CH_3}{\diagdown}} \;+\; HCl \quad \ldots(8)$$

XXXVI   XXXVII

via

$$\left\{ \underset{CH_3}{\overset{CH_3}{\diagdown}}HN:\frown\underset{Cl}{\overset{CH_3}{\underset{\|}{C}}}=O \;\;\longrightarrow\;\; H-\overset{CH_3}{\underset{CH_3}{\overset{|}{N^+}}}-\overset{CH_3}{\underset{Cl}{\overset{|}{C}}}-O^- \;\xrightarrow{-H^+} \right\}$$

In reaction 7, the non-bonding electrons of the oxygen atom in compound XXXV form a two electron bond with H^+ whereby the latter acquires the stable electronic configuration of helium. The oxygen atom retains the octet of electrons it had in I, but it now has only a half share of the pair it has donated to H^+, that is, it is one electron short and now carries the positive charge that was originally associated with the hydrogen atom. The situation is analogous to the formation of the hydroxonium ion H_3O^+ (Unit 9). The subsequent transformations which ultimately lead to the product are similar in nature. In reaction 8 the non-bonding electrons of the nitrogen atom form a two-electron bond to the carbon atom of XXXVII, which is electron deficient due to polarization caused by oxygen being more electronegative than carbon. Two electrons of the double bond are transferred to the oxygen atom simultaneously with the formation of the new bond, thus maintaining an octet of electrons around all the atoms. However, if you count up the electrons associated with the nitrogen and oxygen atoms in the intermediate product XXXVIII, you will find that nitrogen has a half share of four bonding electron pairs and oxygen a half share of one bonding electron pair plus three non-bonding electron pairs. Comparison with the numbers they each had in the neutral compounds XXXVI and XXXVII accounts for the charges in XXXVIII.

Chemical reactions can bring about profound or subtle changes in structure, and through them, large molecules can be built up from or broken down into smaller ones. In all of these processes we can recognize four distinct types of chemical transformation:

    (i) *replacement*, where one atom or group is replaced by another,   replacement

    (ii) *addition*, where the atoms of a reagent are added to a molecule,   addition

(iii) *elimination*, the reverse of addition,

(iv) *rearrangement*, where bonding has changed but no atom has been added to or broken away from the molecule.

elimination

rearrangement

Classify each of the four reactions given below.

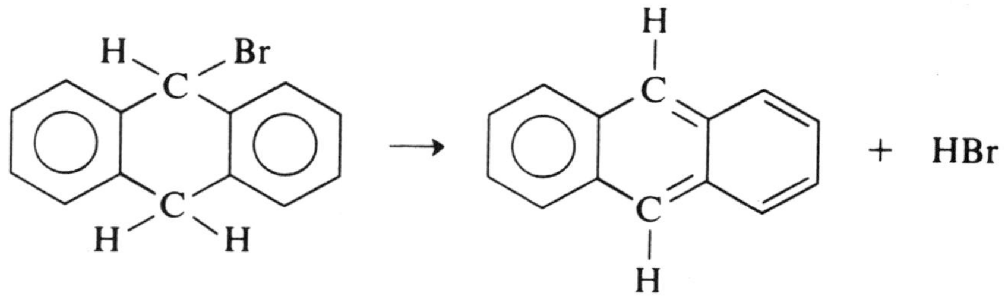

$$\cdots\cdots\cdots\cdots\cdots + HBr \dots\dots\dots\dots\dots\dots\dots\dots (9)$$

$$\frac{H_2}{\longrightarrow} \begin{array}{c} CH_3 \\ \diagdown \\ CH_3 \end{array} CH-OH \dots\dots\dots\dots\dots\dots\dots\dots\dots (10)$$

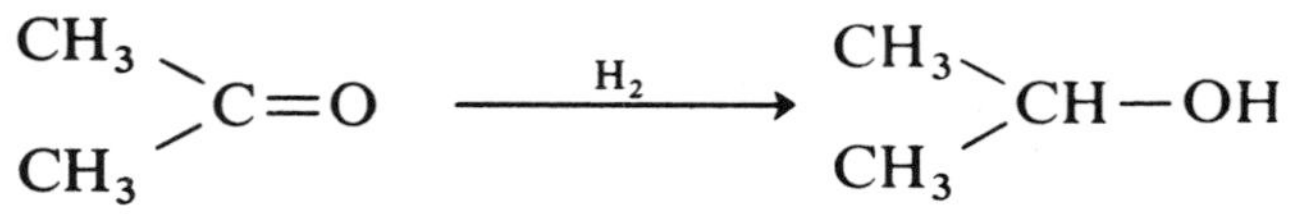

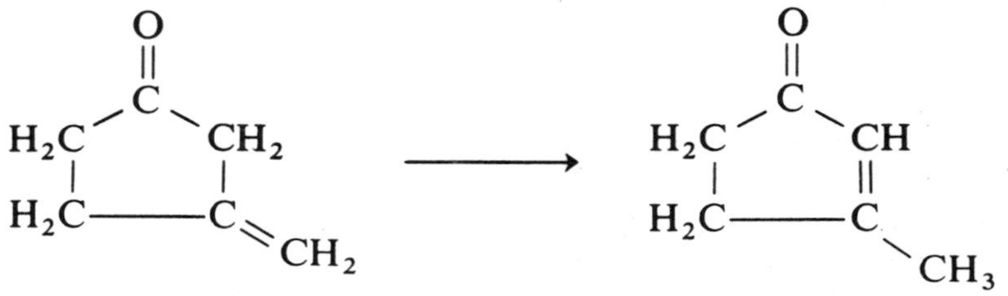

$$\dots\dots\dots\dots\dots\dots\dots\dots (11)$$

$$CH_3.CO.Cl + \cdots\cdots\cdots \longrightarrow \cdots\cdots\cdots + HCl \dots\dots\dots\dots (12)$$

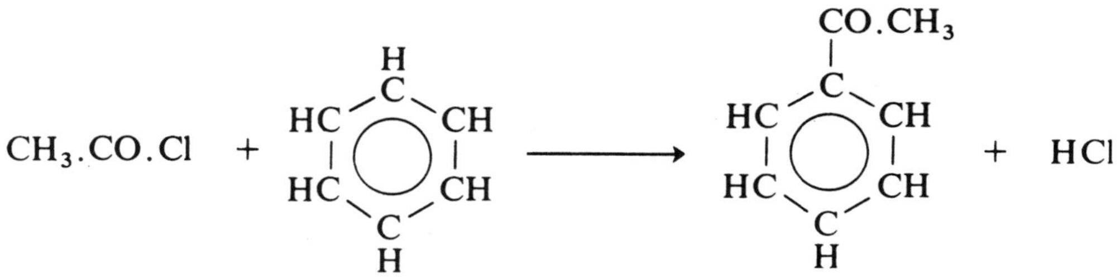

The reactions were those of elimination, addition, rearrangement and replacement respectively. In principle, all four classes of reactions can take place via heterolytic or homolytic routes. We have seen that polar-covalent bonds tend to break so that the more electronegative atom claims both bonding electrons. Such a process can lead to replacement or elimination (see diagram on the following page).

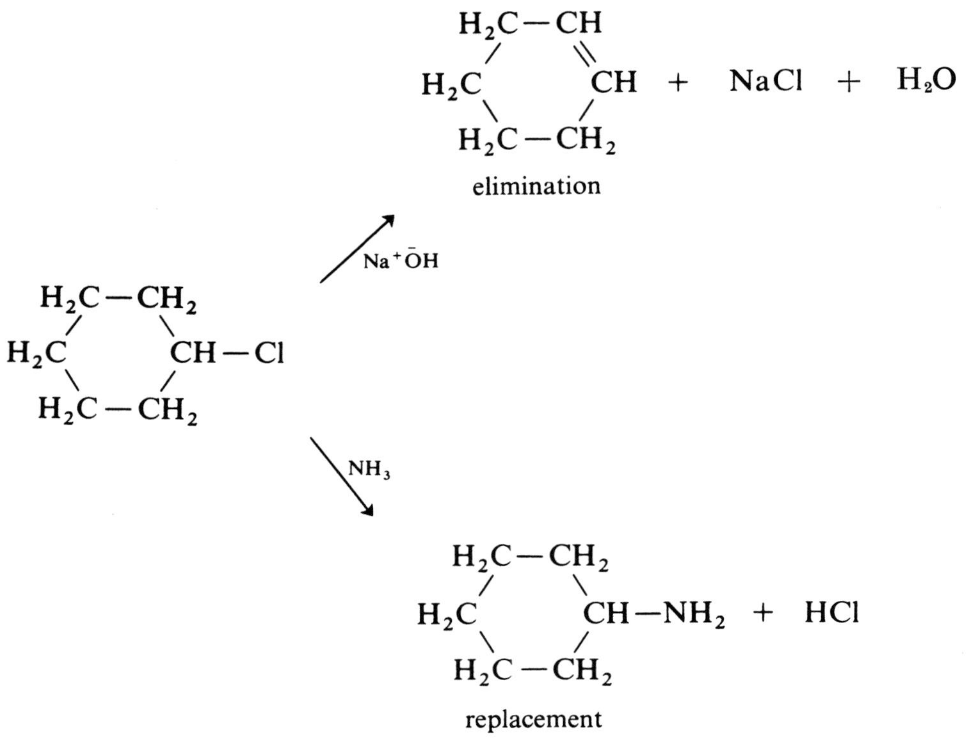

$$\begin{array}{c}
\text{H}_2\text{C}-\text{CH} \\
\diagup \qquad \diagdown\diagdown \\
\text{H}_2\text{C} \qquad\quad \text{CH} \quad + \quad \text{NaCl} \quad + \quad \text{H}_2\text{O} \\
\diagdown \qquad \diagup \\
\text{H}_2\text{C}-\text{CH}_2
\end{array}$$

elimination

$$\begin{array}{c}
\text{H}_2\text{C}-\text{CH}_2 \\
\diagup \qquad\quad \diagdown \\
\text{H}_2\text{C} \qquad\quad \text{CH}-\text{Cl} \\
\diagdown \qquad\quad \diagup \\
\text{H}_2\text{C}-\text{CH}_2
\end{array}$$

$\text{Na}^+\bar{\text{O}}\text{H}$

NH_3

$$\begin{array}{c}
\text{H}_2\text{C}-\text{CH}_2 \\
\diagup \qquad\quad \diagdown \\
\text{H}_2\text{C} \qquad\quad \text{CH}-\text{NH}_2 \quad + \quad \text{HCl} \\
\diagdown \qquad\quad \diagup \\
\text{H}_2\text{C}-\text{CH}_2
\end{array}$$

replacement

Addition reactions are a feature of the chemistry of compounds with multiple bonds, with the notable exception of compounds like benzene, which tend to undergo replacement reactions. It is very difficult to generalize about rearrangements; they are always associated with special structural features within the original molecule. For instance, the driving force for reaction 11 (p. 69) is the tendency for the C=C to become conjugated with the C=O group, a more stable arrangement.

Appendix 7 (Black)

Nomenclature in Organic Chemistry

In writing the text material of Unit 10, we have deliberately limited the number of new chemical names and terms in order to encourage you to focus attention on the concepts rather than the nomenclature. However, one cannot get much further in chemistry without grasping the nettle of terminology, if for no other reason than communication of chemical facts and ideas. There is a limit to the use of such phrases as 'compound 5' or 'structure I' or just plain '1 and 5', and the necessity of drawing structural formulae is avoided if we can refer to compounds by their names.

Most well known organic compounds have at least two names, one of which is often trivial and usually historical in origin, and the other derived from an internationally agreed system of nomenclature. The use of the latter is clearly essential if every organic substance is to have a completely descriptive name which permits only one structural formula to be written for it. Completely undescriptive names should be avoided as far as possible but this is not always practical. Quite obviously 9-(2,6,6-trimethyl-1-cyclohexenyl)-3,7-dimethyl-2,4,6,8-nona-tetraen-1-ol has phonetic disadvantages as a handy name for vitamin A!

The limited time you have available for reading this appendix restricts our present treatment of nomenclature to one of setting out the names of the commonest groups found in organic compounds and some ways these can be handled to produce satisfactory names for complete molecular formulae. By and large, organic compounds can be considered as carbon frameworks to which are added so-called functional groups; that is, groups where bonding is relatively easily broken.

$$(C_nH_{2n+2})$$

CH_4	methane	C_6H_{14}	hexane
C_2H_6	ethane	C_7H_{16}	heptane
C_3H_8	propane	C_8H_{18}	octane
C_4H_{10}	butane	C_9H_{20}	nonane
C_5H_{12}	pentane	$C_{10}H_{22}$	decane
		$C_{14}H_{30}$	tetradecane . . . etc.

The names of hydrocarbon groups derived from such compounds (alkyl groups) are obtained by replacing the suffix -ane by -yl. Thus we have CH_3-methyl, C_3H_7-propyl, etc. You saw in the text that saturated hydrocarbons with four or more carbon atoms can exist as any one of a number of structural isomers. Branched chain hydrocarbons are named in the following way. The longest continuous chain is identified and the compound is regarded as a derivative of the corresponding straight chain alkane.

$$\overset{1}{CH_3}-\overset{2}{CH}-\overset{3}{CH_2}-\overset{4}{CH}-CH_3$$
$$\quad\quad |\quad\quad\quad\quad\quad |$$
$$\quad\quad CH_3\quad\quad\quad CH_2-CH_3 \quad XXXIX$$
$$\quad\quad\quad\quad\quad\quad\quad\quad 5\quad\quad 6$$

Thus the molecule XXXIX shown would be a derivative of hexane. The carbon atoms of the chain are then numbered from the end nearest to branching and the substituents, other than hydrogen atoms, named and numbered. Thus the compound XXXIX becomes 2,4-dimethylhexane (not 2-methyl-4-ethylpentane).

Some of the small branched chain hydrocarbon groups are most frequently referred to by another system.

$CH_3CH_2CH_2$ n-propyl n = normal

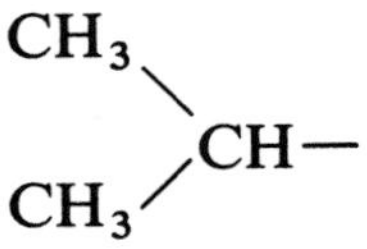

$$\begin{array}{c} CH_3 \\ \diagdown \\ \quad\quad CH- \\ \diagup \\ CH_3 \end{array}$$ iso-propyl

$CH_3CH_2CH_2CH_2-$ n-butyl

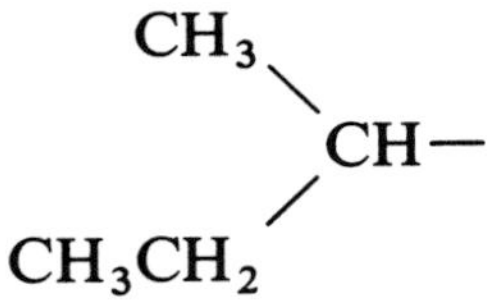

$$\begin{array}{c} CH_3 \\ \diagdown \\ \quad\quad CH- \\ \diagup \\ CH_3CH_2 \end{array}$$ sec-butyl sec = secondary

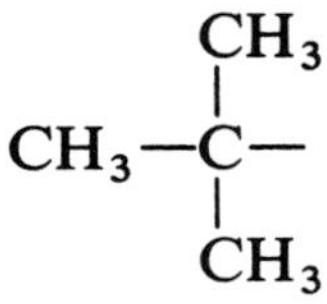

$$\begin{array}{c} CH_3 \\ | \\ CH_3-C- \\ | \\ CH_3 \end{array}$$ t-butyl t = tertiary

Monocyclic compounds are named after their straight chain counterparts with the prefix *cyclo* added. Compound XL is thus

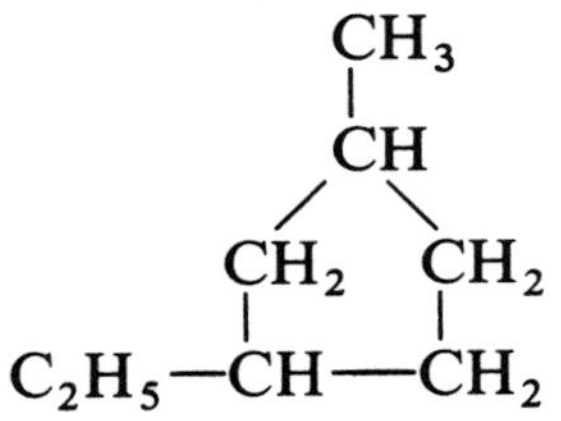

$$\begin{array}{c} CH_3 \\ | \\ CH \\ \diagup\quad\diagdown \\ CH_2\quad CH_2 \\ |\quad\quad| \\ C_2H_5-CH-CH_2 \end{array}$$ XL

1-methyl-3-ethylcyclopentane.

The system for polycyclic saturated hydrocarbons is rather complex.

When double bonds are involved in hydrocarbon structures (alkenes), the ending *-ene* is used instead of *-ane*. The numbering and naming then follows logically. Thus structure XLI becomes 4-methyl-2-pentene. The

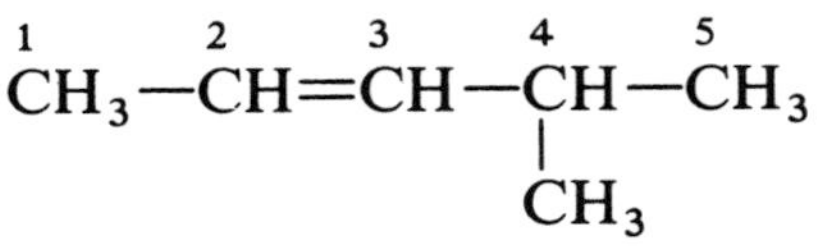

$$\overset{1}{C}H_3-\overset{2}{C}H=\overset{3}{C}H-\overset{4}{C}H-\overset{5}{C}H_3$$
$$| $$
$$CH_3$$ XLI

$$\begin{array}{c} H \\ \diagdown\quad\quad\diagup \\ \quad C=C \\ \diagup\quad\quad\diagdown \\ \quad\quad\quad\quad H \end{array}$$
XLII   *trans*

$$\begin{array}{c} \diagdown\quad\quad\diagup \\ \quad C=C \\ \diagup\quad\quad\diagdown \\ H\quad\quad\quad H \end{array}$$ *cis*

prefixes *cis* and *trans* will apply according to the geometry about the double bond, XLII (10.4.3).

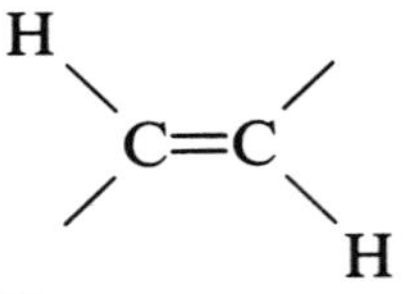

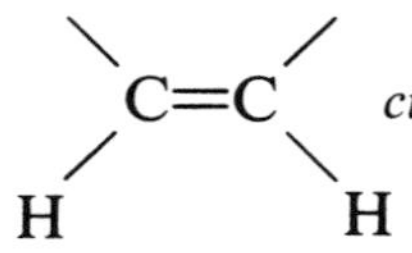

Compounds containing carbon triple bonds are known as alkynes and their names are derived as above, the suffix *-ane* being replaced by *-yne*.

Appendix 7

Thus structure XLIII becomes hex-1-ene-5-yne.

$$CH_2{=}CH{-}CH_2{-}CH_2{-}C{\equiv}CH \quad \text{XLIII}$$

As might be expected, derivatives of benzene have a nomenclature system of their own. The ring carbon atoms are numbered from 1 to 6, the lowest number going to the carbon atom bearing the most important substituent. Thus

XLIV

XLV

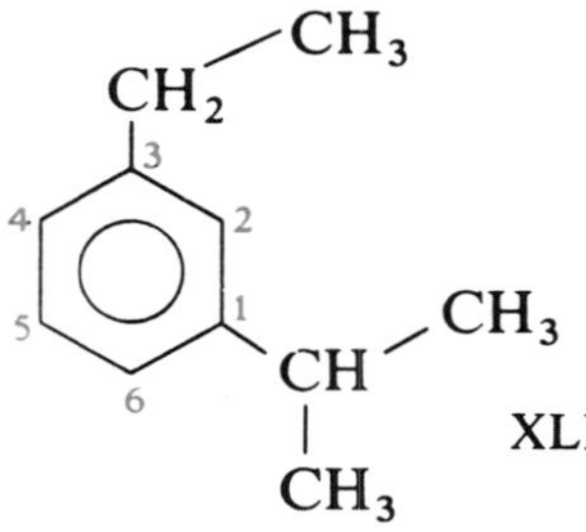

XLIV becomes 3-ethyl*iso*propylbenzene. The group XLV is known as phenyl (Ph.). Compound XLVI

XLVI

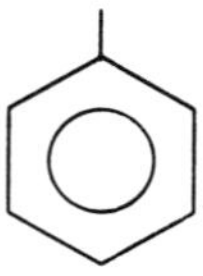

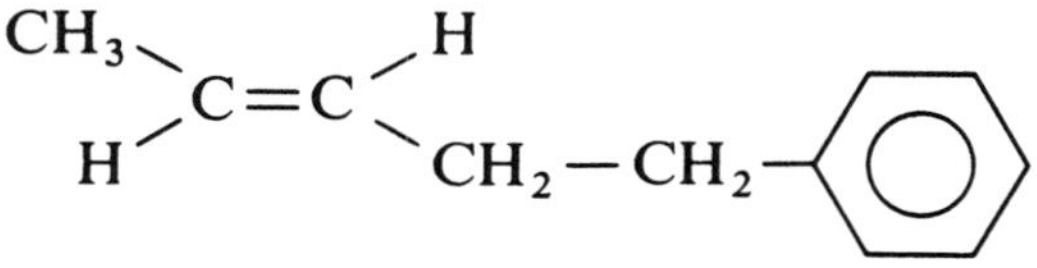

would be known as 5-phenyl-*trans*-pent-2-ene.

The major functional groups are set out in Table 6, p. 74. There are essentially two ways of referring to the groups. They can be used to classify an organic compound or they can be referred to as a group substituent.

Recommended parallel reading

Nuffield Advanced Science: Chemistry, Book I, Topic 9 Appendix.

Table 6

Names of some common Functional Groups

Group	Class of compounds which contain the group	Variants for benzenoid compounds	Name of group when regarded as a substituent	Example
saturated hydrocarbon	alkanes	methyl benzene = toluene	alkyl	CH_4 methane
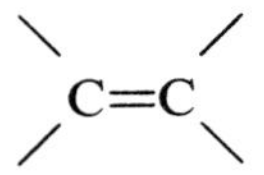 $\diagdown C = C \diagup$	alkenes		alkenyl (CH_2=CH— = vinyl)	$CH_2 = CH—CH_3$ propene
—C≡C—	alkynes		alkynyl	$CH_3—C≡CH$ propyne
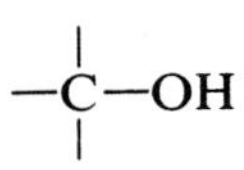 —C—OH	alkanols (alcohols)	hydroxybenzene = phenol	hydroxy	$CH_3—CH_2—OH$ ethanol
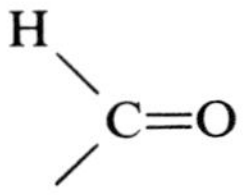 H $\diagdown C = O \diagup$	alkanals (aldehydes)	phenylcarboxaldehyde = benzaldehyde	carboxaldehyde	$CH_3—CHO$ ethanal (acetaldehyde)
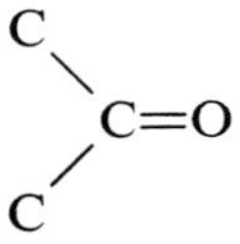 C $\diagdown C = O$ C	alkanones (ketones)		oxo	$CH_3—CO—C_2H_5$ butan-2-one
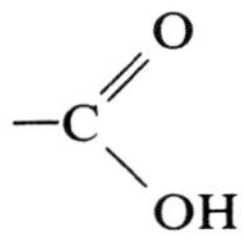 —C⟨O / OH	alkanoic acids	phenylcarboxylic acid = benzoic acid	carboxylic acid	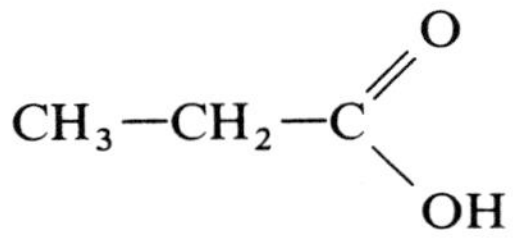 $CH_3—CH_2—C⟨O/OH$ propionic acid
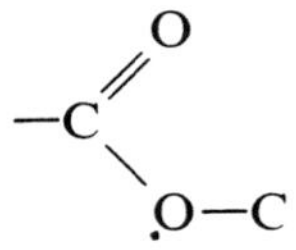 —C⟨O / O—C	carboxylic esters		carboalkoxy	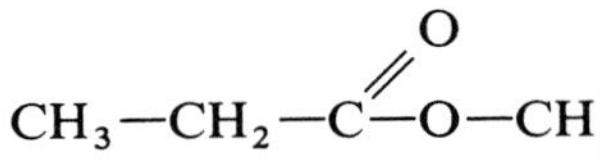 $CH_3—CH_2—C⟨O/O—CH$ carbomethoxyethane (methyl propionate)
C—O—C	ether		alkoxy	$CH_3—CH_2—O—CH_3$ methoxyethane (ethylmethylether)
C—Hal	alkyl halides		halogeno (fluoro, chloro, bromo, iodo)	$CH_2 = CCl_2$ 1,1′-dichloroethylene
C—NH_2	alkylamines	aminobenzene = aniline	amino	$CH_3CH_2—NH_2$ ethylamine (1-amino propane)

Self-Assessment Questions

Sections 10.1 and 10.2

Question 1 *(Objective 1)*

Tick to show which of the expressions describing chemical compounds listed on the left correspond to which of the terms below.

	(a) covalent compounds	(b) ionic compounds	(c) hydro-carbons	(d) structural isomers
(i) composed of hydrogen and carbon atoms only				
(ii) bonding is largely electro-static				
(iii) composed of the same number of atoms of different elements but are nevertheless different substances				
(iv) electrons are shared between atoms				

Question 2 *(Objective 1)*

Tick to show which of the following statements are true, and which false.

	(a) true	(b) false
(i) The carbon hydrogen bond in a hydrocarbon owes its strength partly to the fact that hydrogen acquires the electronic configuration of neon and carbon the electronic configuration of helium		
(ii) The valence electrons of an atom are to be found in its highest occupied shell		
(iii) The carbon-carbon bonds in hydrocarbons are highly polar		
(iv) Open chain isomers of mono-cyclic hydro-carbons with one double bond, will always contain two double bonds or a triple bond		

Question 3 *(Objective 3)*

Tick to show the names of the structures below.

	(a)	(b)	(c)	(d)	(e)
	$CH_2{=}CH_2$	$CH_3{-}CH_2{-}CH_3$	CH_4	$CH{\equiv}CH$	$CH_3{-}CH_3$

(i) acetylene

(ii) ethane

(iii) ethylene

(iv) propane

(v) methane

Section 10.3

Question 4 *(Objective 4)*

Place a tick against the two dimensional structures below which you think represent viable organic molecules.

(i) (   )   (ii) 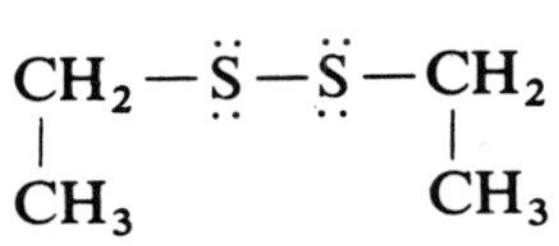 (   )   (iii) (   )

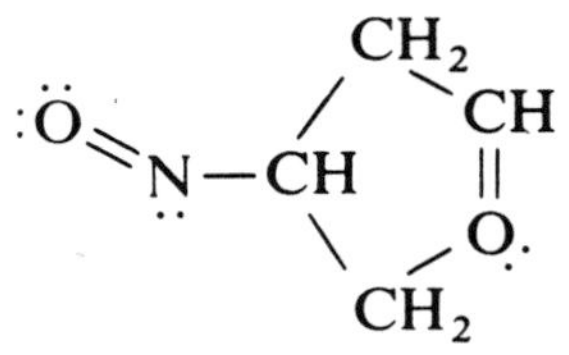

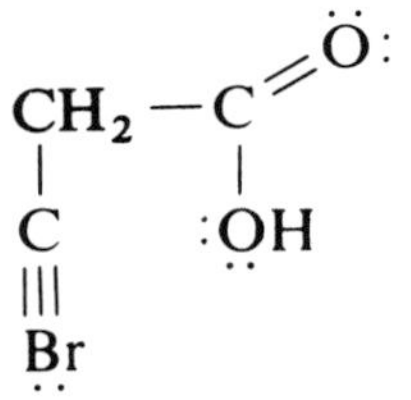

(iv) 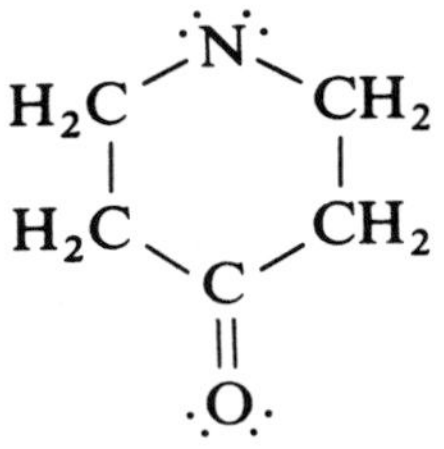 (   )   (v) 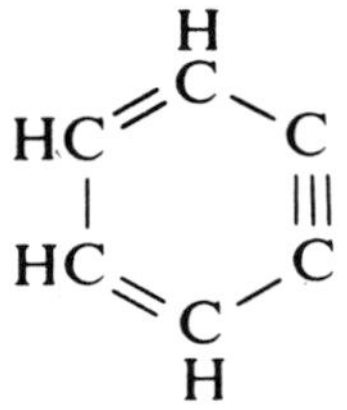 (   )   (vi) (   )

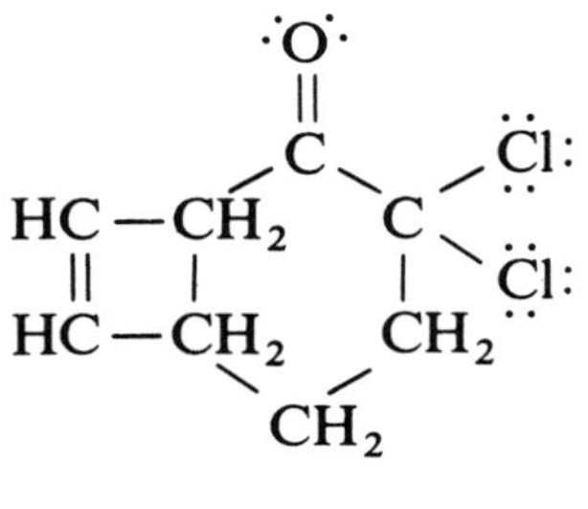

Question 5 *(Objective 6)*

Look at the silicon analogue of the carbon compound given below.

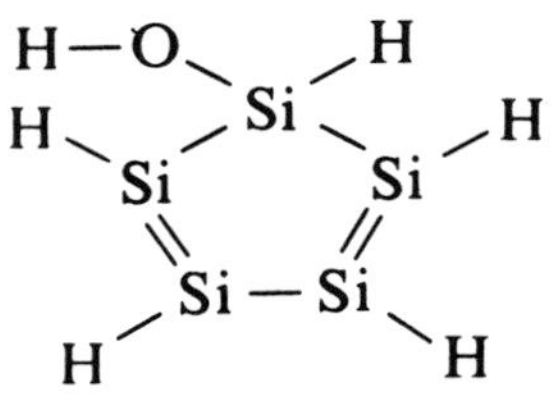

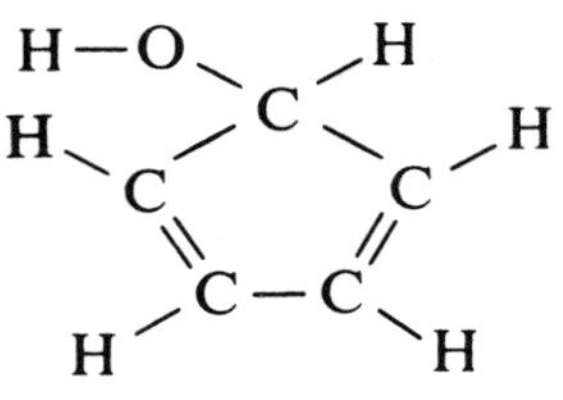

For which of the reasons given below would you expect the silicon analogue to be unstable.

(i) Silicon does not form multiple bonds very readily

(ii) Compounds of silicon always have open-chain structures

(iii) The silicon oxygen bond is much weaker than the carbon oxygen bond

(iv) Silicon silicon bonds are relatively weak

(v) The silicon hydrogen bond is chemically reactive

Question 6 *(Objective 8)*

Arrange the following reasons in order of importance to the question: Why does carbon form more covalent compounds than all the other elements combined?

 (i) carbon forms strong bonds with many elements

 (ii) carbon is tetravalent

(iii) the carbon–carbon bond is very strong

(iv) elemental carbon is stable in air

(The following three items are 'short answer' questions. Write your answers on a separate sheet of paper).

Question 7 *(Objective 6)*

Give some reasons (in less than three hundred words) why the compounds of carbon and silicon play different roles in nature.

Question 8 *(Objective 9)*

Why is the chemistry of carbon compounds called organic chemistry? Use no more than fifty words.

Section 10.4

Question 9 *(Objective 11)*

State briefly the influence on bond angles in a molecular fragment if

(a) bonding electron pairs are replaced by non-bonding electron pairs

(b) one of the bonds is a double bond

(c) there is a large difference in electronegativity between the bonded atoms.

Question 10 *(Objective 1)*

Tick to show which of the conclusions (a) to (e) about the statements below is correct.

(a) Both assertion and reason are true statements and the reason is a correct explanation of the assertion

(b) Both assertion and reason are true statements but the reason is NOT a correct explanation of the assertion

(c) The assertion is true but the reason is a false statement

(d) The assertion is false but the reason is a true statement

(e) Both assertion and reason are false statements

                     (a)   (b)   (c)   (d)   (e)

(i) *Assertion*
Chirality can be demonstrated with the molecule X

Reason
It does not possess a plane of symmetry

$$H-\overset{\overset{\displaystyle OH}{|}}{\underset{\underset{\displaystyle CH_3}{|}}{C}}-\overset{\overset{\displaystyle O}{\|}}{C}-CH_3$$

molecule X

(ii) *Assertion*
Molecule X is saturated

Reason
It contains atoms with non-bonding electrons

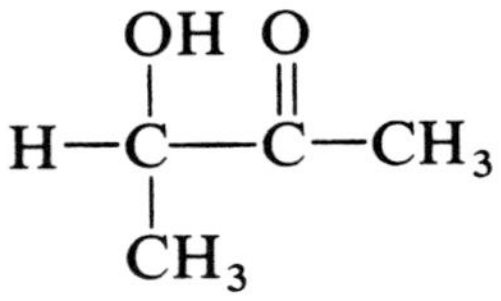

Question 11 *(Objective 10)*

The bond angle in carbon dioxide, CO_2 is expected to be

(i) 90°

(ii) 120°

(iii) 180°

(iv) none of the above

Question 12 *(Objective 11)*

Which of the three possibilities below represents the order in which the inter-electron repulsions in nitrosyl fluoride, $N=O$, decrease

$$\underset{F}{\nearrow}$$

(i) $\cdot\cdot, N=O;\quad N=O, N-F;$ ( )
     $\cdot\cdot, N-F$

(ii) $\cdot\cdot, N-F;\quad \cdot\cdot, N=O;$ ( )
     $N=O, N-F$

(iii) $\cdot\cdot, N-F;\quad N=O, N-F;$ ( )
     $\cdot\cdot, N=O$

$\cdot\cdot$—non bonding electrons

$N=O$—electrons of double bond

$N-F$—electrons of single bond

Question 13 *(Objective 12)*

Tick to show which of the following do, and which do not, represent viable 3-D structures of organic compounds.

| directed up from paper

⋮ directed down behind paper

		(a) viable	(b) not viable
(i)	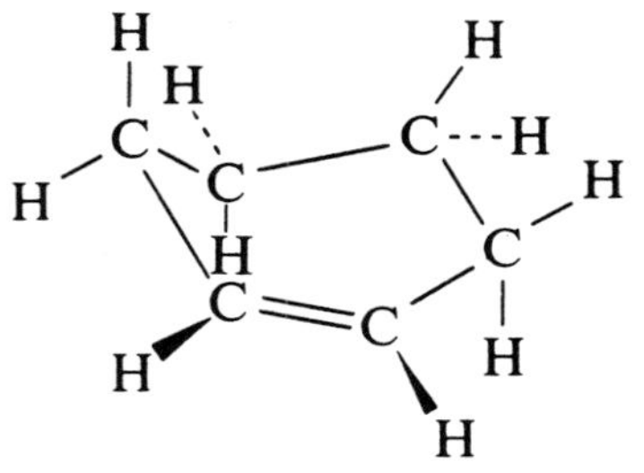	( )	( )
(ii)	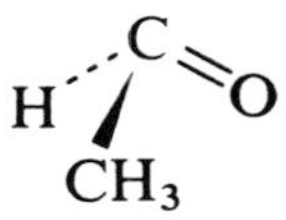	( )	( )
(iii)	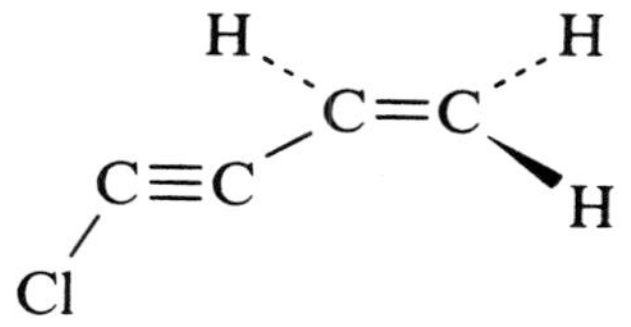	( )	( )
(iv)	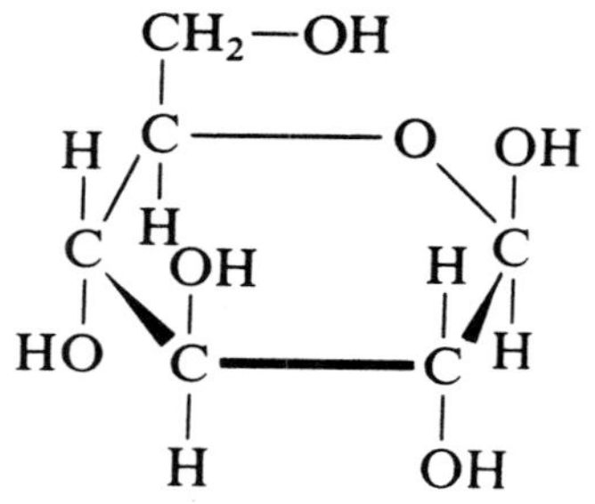	( )	( )
(v)	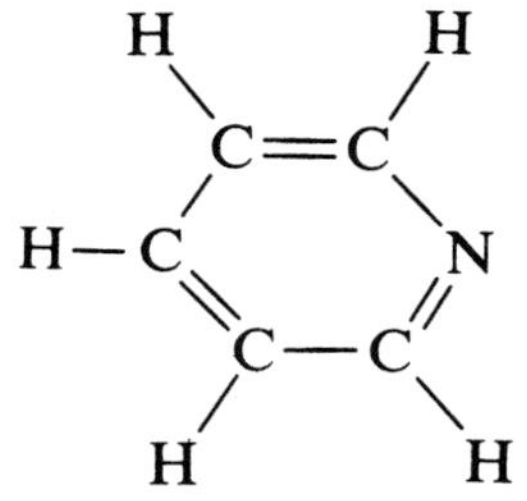	( )	( )
(vi)	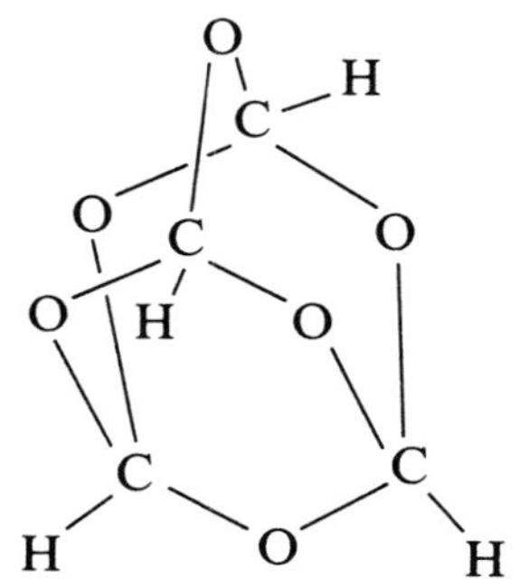	( )	( )

Question 14   *(Objective 15)*

Tick to show which of the structures below represents an optical isomer, and which a geometric isomer, of the structure A.

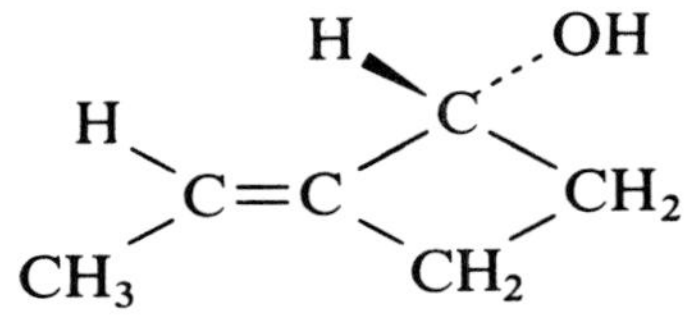

		(a) optical isomer	(b) geometric isomer	(c) neither
(i)	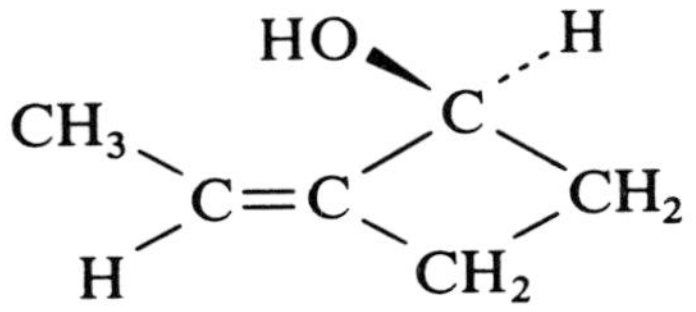	( )	( )	( )
(ii)	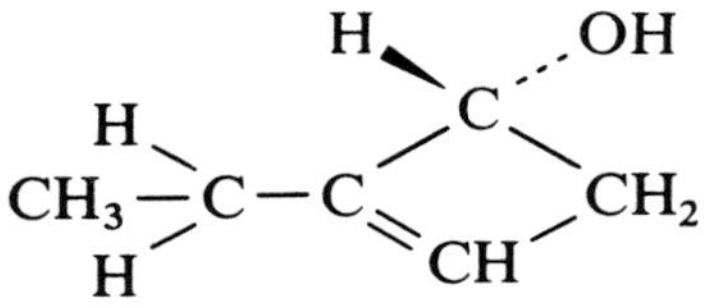	( )	( )	( )
(iii)	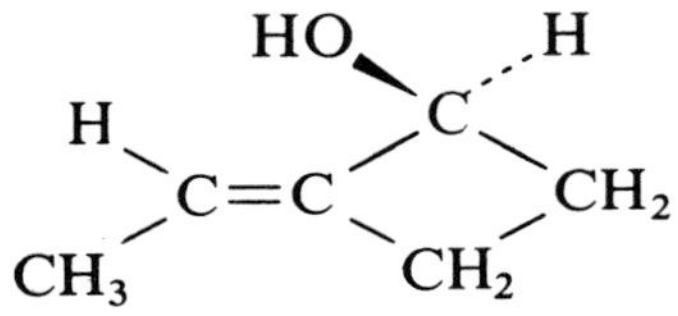	( )	( )	( )
(iv)	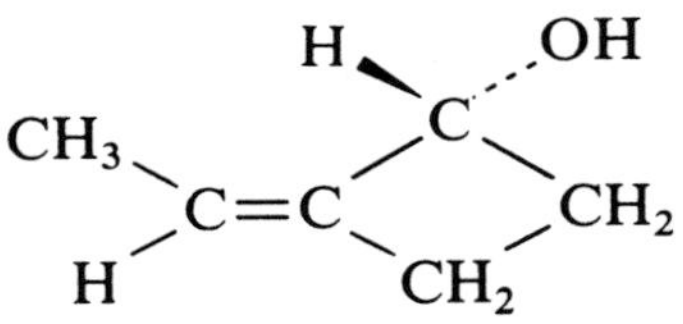	( )	( )	( )

Question 15 *(Objectives 1, 15, 17, 18, 19)*

Place ticks in the table below to show which of the features and properties listed on the left are applicable to each of the compounds at the top. (Some features may not be applicable to any.)

compound feature/property	(a) alanine	(b) tetra- hydro- furan	(c) glucose	(d) glycine	(e) mannose	(f) diethyl ether
(i) chiral	( )	( )	( )	( )	( )	( )
(ii) unsaturated	( )	( )	( )	( )	( )	( )
(iii) possesses a ring	( )	( )	( )	( )	( )	( )
(iv) carbo-hydrate	( )	( )	( )	( )	( )	( )
(v) aminoacid	( )	( )	( )	( )	( )	( )
(vi) stereo-isomer present in this table	( )	( )	( )	( )	( )	( )
(vii) saturated	( )	( )	( )	( )	( )	( )
(viii) can readily adopt a shape which possesses a plane of symmetry	( )	( )	( )	( )	( )	( )
(ix) can exist in one con-formation only	( )	( )	( )	( )	( )	( )
(x) achiral	( )	( )	( )	( )	( )	( )
(xi) 'building block' of cellulose	( )	( )	( )	( )	( )	( )
(xii) can exist in 'chair' and 'boat' forms	( )	( )	( )	( )	( )	( )
(xiii) could be a 'building block' of a protein	( )	( )	( )	( )	( )	( )
(xiv) occurs naturally as the (+) isomer	( )	( )	( )	( )	( )	( )
(xv) occurs naturally as the (−) isomer	( )	( )	( )	( )	( )	( )

81

Question 16 *(Objective 14)*

Draw a three dimensional structure for glucose.

Section 5

Question 17 *(Objectives 1, 13, 19)*

Carbon exists in two crystalline forms, diamond and graphite. In both cases, carbon atoms are covalently bound to one another and yet the two forms have profoundly different properties. One is an extremely hard material of rigid structure and without colour. The other is very soft, black, and is used as a lubricant and in pencil 'lead'. The array of carbon atoms in each crystalline form is shown below.

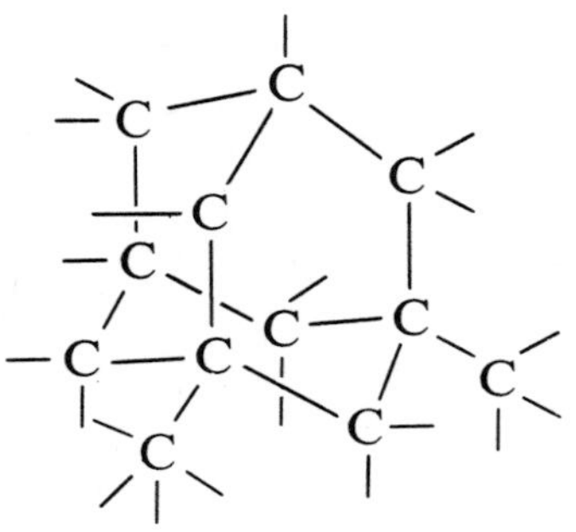

I

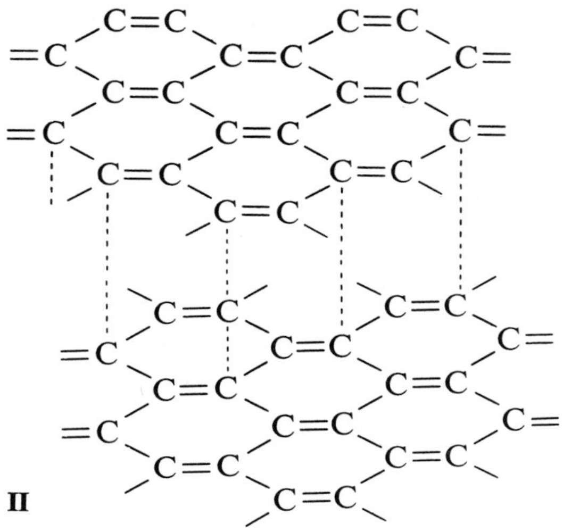

II

Recall from Unit 8 which structure applies to which compound and relate structural features to the properties given above, and using the concepts of bonding and polarizability.

(Write your answer in about 300 words on a separate sheet of paper).

Question 18 *(Objective 19)*

Arrange the following compounds in order of increasing boiling point or melting point.

(i) (a)

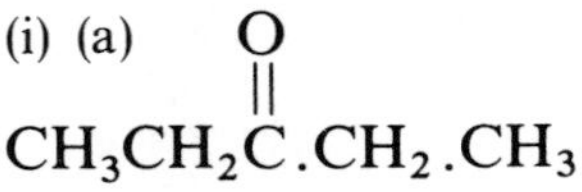

$$CH_3CH_2\overset{\overset{\displaystyle O}{\|}}{C}.CH_2.CH_3$$

(b)

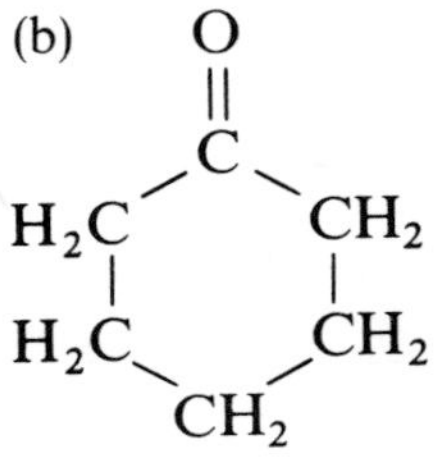

(c)

$$CH_3\overset{\overset{\displaystyle O}{\|}}{C}CH_3$$

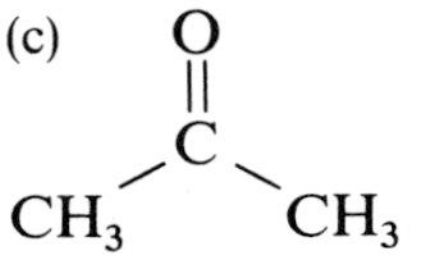

(d)

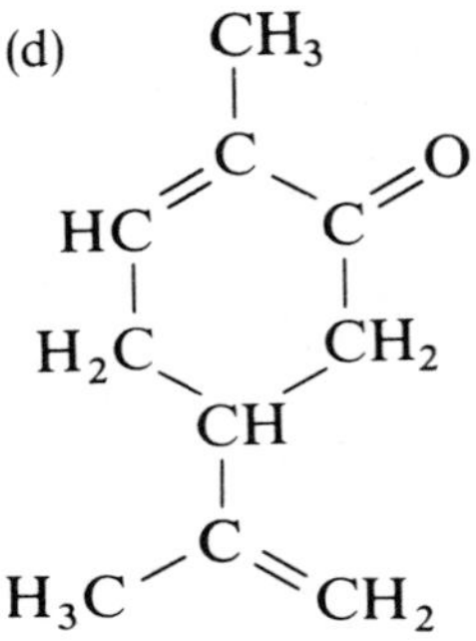

(ii) (a)

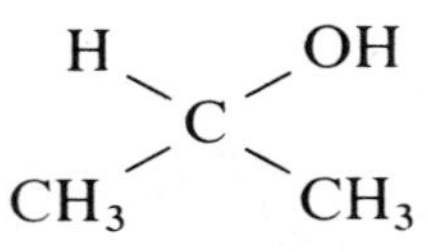

$$H \diagdown \underset{C}{\diagup} OH$$
$$CH_3 \diagup \diagdown CH_3$$

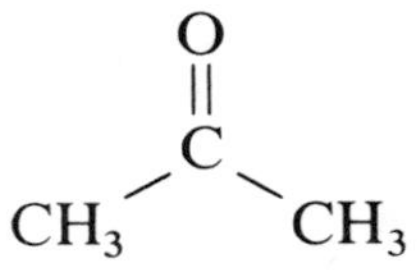

(b)

$$O$$
$$\|$$
$$C$$
$$CH_3 \diagup \diagdown CH_3$$

(c)

$$CH_2$$
$$\|$$
$$C$$
$$CH_3 \diagup \diagdown CH_3$$

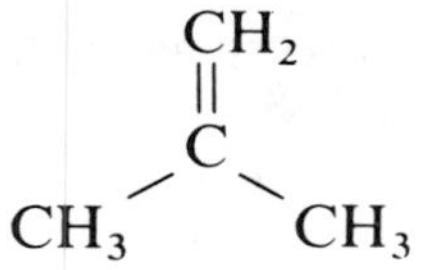

Question 19 (Objective 19)

Arrange the following compounds in order of increasing solubility in water.

(i) (a)

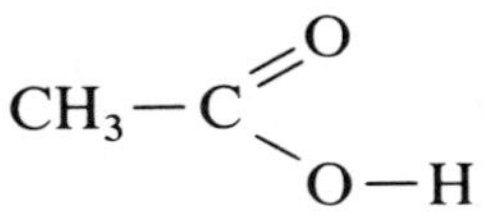

$$CH_3 - C \diagup^{\displaystyle O}_{\displaystyle O-H}$$

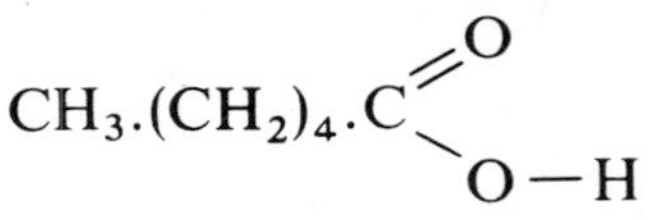

(b)

$$CH_3.(CH_2)_4.C \diagup^{\displaystyle O}_{\displaystyle O-H}$$

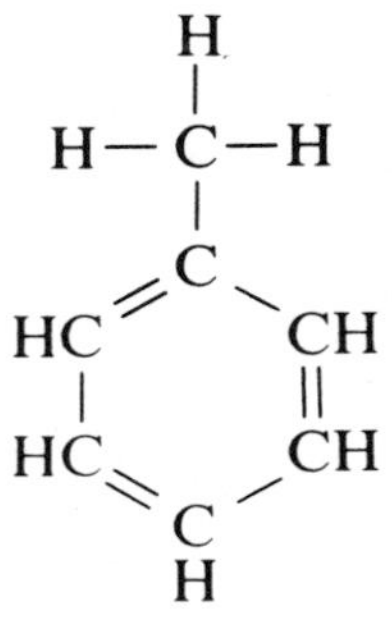

(c)

$$CH_3.(CH_2)_{16}.C \diagup^{\displaystyle O}_{\displaystyle O-H}$$

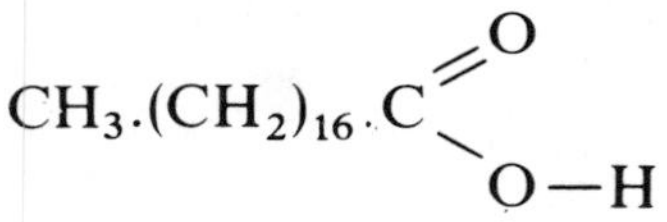

(ii) (a)

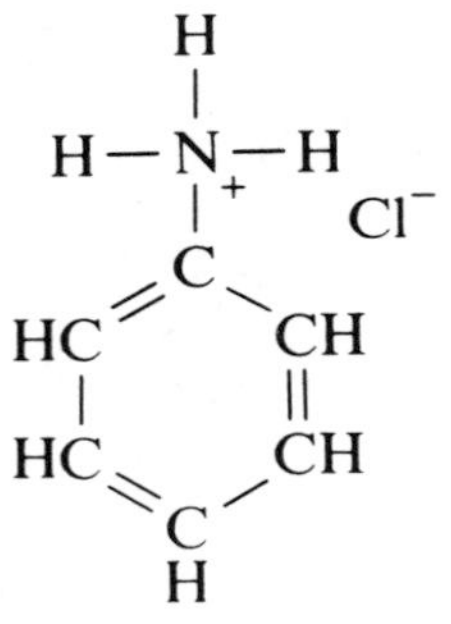

$$
\begin{array}{c}
H \\
| \\
H-N-H \\
| \!\!+ \quad Cl^- \\
C \\
HC \diagup \diagdown CH \\
HC \diagdown \diagup CH \\
C \\
H
\end{array}
$$

(b)

$$
\begin{array}{c}
H \\
| \\
H-C-H \\
| \\
C \\
HC \diagup \diagdown CH \\
HC \diagdown \diagup CH \\
C \\
H
\end{array}
$$

(c)

$$
\begin{array}{c}
H \\
| \\
O \diagdown \diagup O \\
C \\
| \\
C \\
HC \diagup \diagdown CH \\
HC \diagdown \diagup CH \\
C \\
H
\end{array}
$$

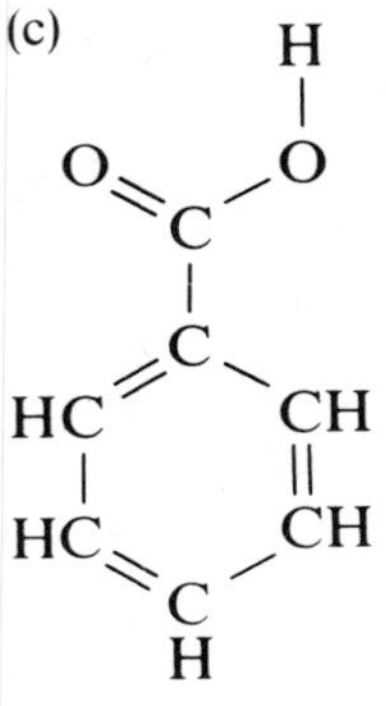

Question 20 (Objective 21)

Distinguish between the following activities according to whether or not they remind you of the reason why an enzyme normally accepts only one of a mirror-image related pair of molecules for catalytic chemical change.

	(a) analogous	(b) not analogous
(i) riding astride a horse	( )	( )
(ii) riding a horse side-saddle	( )	( )
(iii) driving a British car on the Continent	( )	( )
(iv) ballroom dancing	( )	( )

Question 21 *(Objective 20)*

Group the following compounds according to those which might be expected to exhibit similar chemical behaviour.

(i) $CH_3.OH$

(ii) 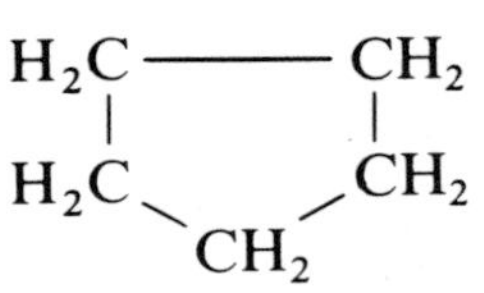 H_2C———CH_2
H_2C CH_2
          CH_2

(iii) 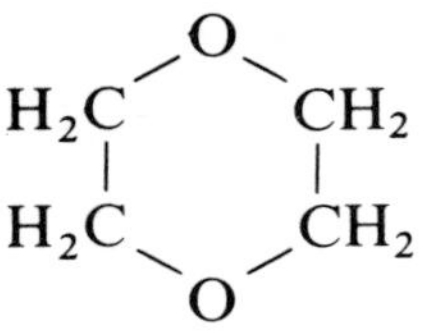 H_2C O CH_2
H_2C CH_2
          O

(iv)

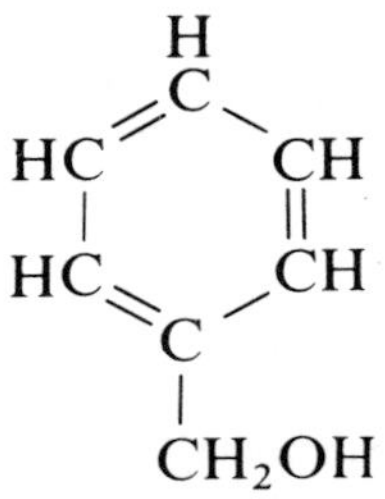

(v) $CH_3.O.CH_3$

(vi) C_5H_{12}

(vii) $CH_2{=}CH_2$

(viii)
HC CH_2
‖ CH_2
HC CH_2

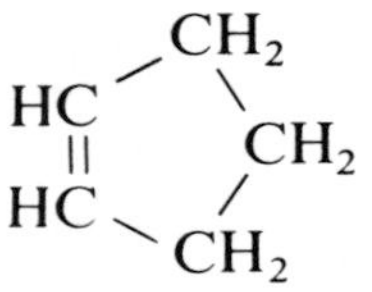

(ix) 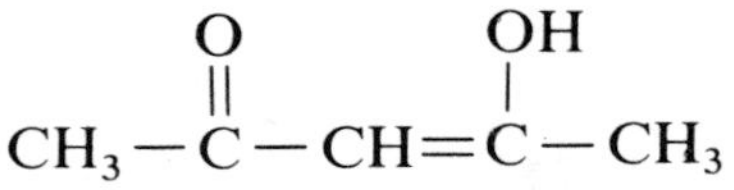
$CH_3{-}\overset{\displaystyle O}{\underset{\displaystyle \|}{C}}{-}CH{=}\overset{\displaystyle OH}{\underset{\displaystyle |}{C}}{-}CH_3$

(x) 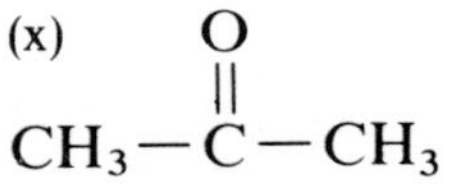
$CH_3{-}\overset{\displaystyle O}{\underset{\displaystyle \|}{C}}{-}CH_3$

(xi)

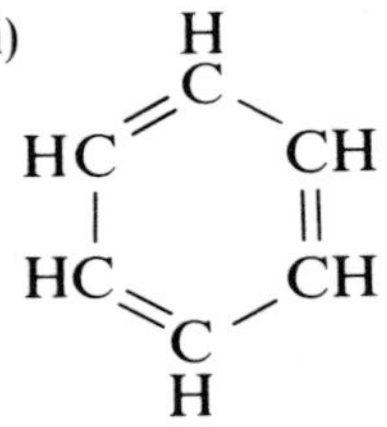

Sections 1, 4, 5

Question 22 *(Objective 1)*

Indicate by a tick in the appropriate column whether each description on the left can be associated with the electronic property of polarization or polarizability.

	(a) polarization	(b) polarizability
(i) bonds whose electrons are responsible for the absorption of visible light	( )	( )
(ii) bonds which have a symmetrical distribution of electrons but are nevertheless relatively easily broken	( )	( )
(iii) *intra*molecular bonding which induces strong *inter*molecular cohesive forces.	( )	( )

Section 6

Question 23 *(Objective 22)*

Look at the structures (i)–(vi) shown below

(i)

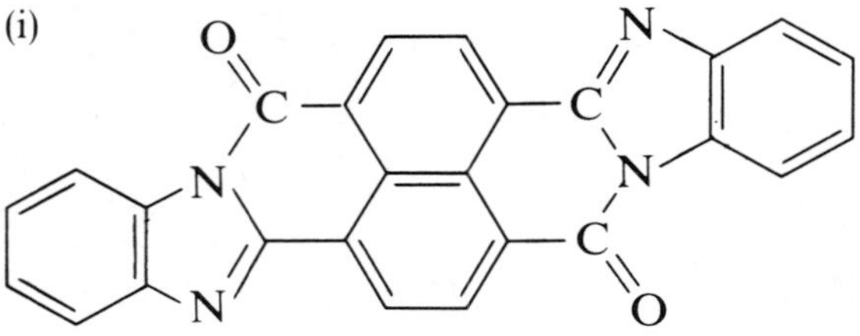

(ii)

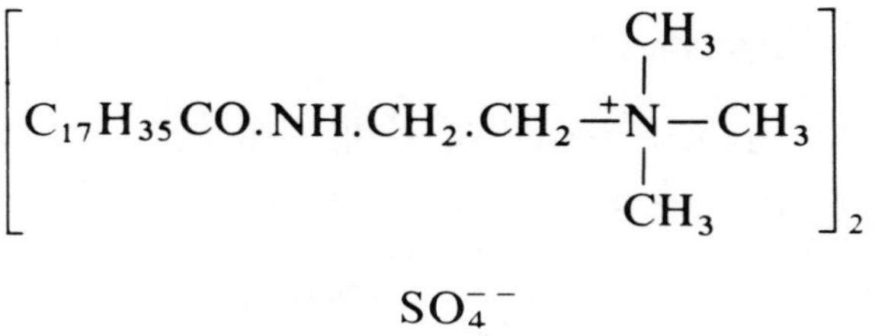

$$\left[C_{17}H_{35}CO.NH.CH_2.CH_2 \overset{+}{-} \underset{\underset{CH_3}{|}}{\overset{\overset{CH_3}{|}}{N}} - CH_3 \right]_2$$

$$SO_4^{--}$$

(iii)

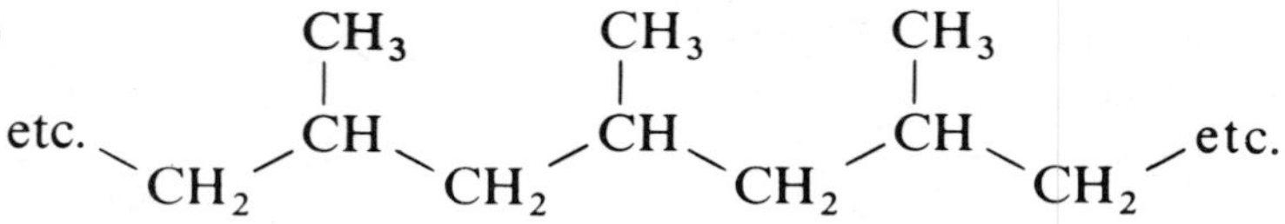

(iv) CH_4

(v)

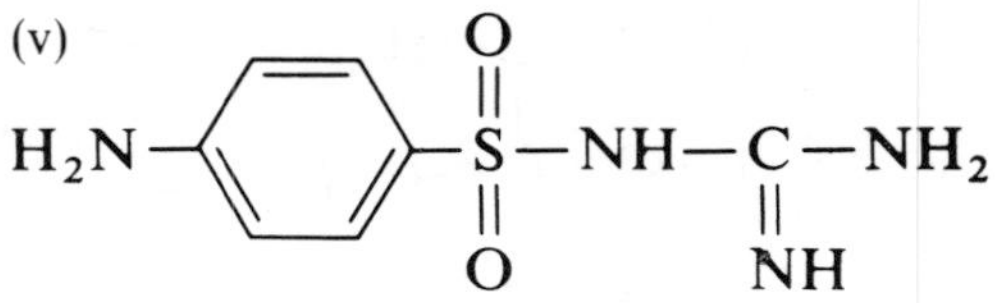

(vi)

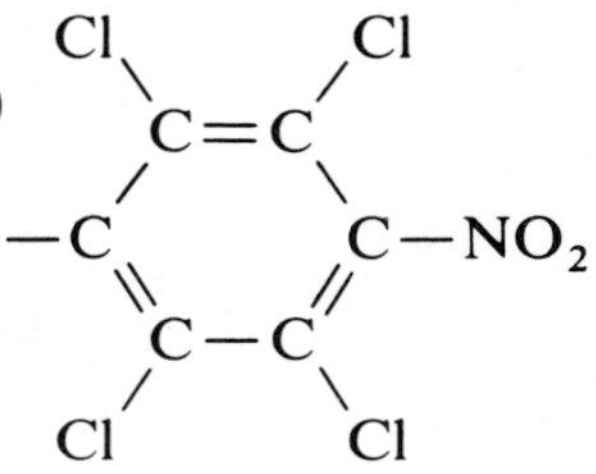

Assign each structure to one of the fields of chemical technology (a)–(f)

	(a) petroleum and petro-chemicals	(b) dyes	(c) pharma-ceuticals	(d) pesticides	(e) polymers	(f) detergents
(i)	( )	( )	( )	( )	( )	( )
(ii)	( )	( )	( )	( )	( )	( )
(iii)	( )	( )	( )	( )	( )	( )
(iv)	( )	( )	( )	( )	( )	( )
(v)	( )	( )	( )	( )	( )	( )
(vi)	( )	( )	( )	( )	( )	( )

Question 24 *(Objective 23)*

List three examples of molecular design in chemical technology

(i)

(ii)

(iii)

Self Assessment Answers
and Comments

Question 1

Answer   (i) c ; (ii) b ; (iii) d ; (iv) a.

Comment   You are referred to your text, sections 1 and 2.

Question 2

Answer   (i) b ; (ii) a ; (iii) b ; (iv) a.

Comment   (i) hydrogen acquires the electronic configuration of helium and carbon that of neon

(iii) polar bonds occur between atoms of different electro-negativity

(iv) You could have deduced this was true by trial and error. The molecular formula of a compound is affected equally by the presence of a double bond and a ring.

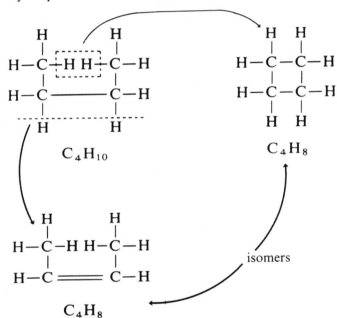

See also sections 1 and 2.

Question 3

Answer   (i) d ; (ii) e ; (iii) a ; (iv) b ; (v) c.

Comment   See sections 1 and 2.

Question 4

Answer   (ii) (v).

Comment   (i) The ring oxygen atom has one non-bonding pair and a half share of three bonding pairs giving a total of five valence electrons. Oxygen has six.

(iii) The bromine atom has one non-bonding pair and a half share of a six-electron bond giving a total of five valence electrons. Bromine has seven.

(iv) The ring nitrogen atom has two non-bonding pairs
and a half share in two bonding pairs giving a total of
six valence electrons. Nitrogen has five.

(vi) The carbon atoms common to both rings are each bound
to *five* other atoms, whereas the maximum permissible for
carbon is four.

See also section 3.

Question 5

Answer   (i), (iv), (v).

Comment   (ii) There is nothing to prevent the $-Si-O-Si-O$
chains linking around to form rings.

(iii) Quite the reverse — see your text (section 3).

Question 6

Answer   (iii), (i), (ii), (iv).

Comment   (iv) Is not directly relevant.

(ii) Other things being equal, carbon could have formed a
tremendous variety of structures had it been trivalent.
The enormous variety in carbon skeletons possible
through (iii) must be the most important factor here.
See also section 3.

Question 7

Answer   The Si—Si bond is weak whereas the C—C bond is strong.
Silicon does not readily participate in multiple bonds whereas
carbon does. The C—O bond though strong is much weaker
than the Si—O bond. Silicon occurs naturally in compounds
composed largely of silicon oxygen bonds. Large single bonded
chain-like structures are common and constitute much of the
material found in sands, rocks, soils, etc.

In addition to the capability of forming a variety of structures
bonded to other carbon atoms through single and multiple
bonds, carbon forms strong bonds to many other elements
especially hydrogen, partly because of its electronegativity
value which results from its electronic configuration. The
consequent enormous variety in structure and hence, behaviour
of carbon compounds is largely responsible for its role as the
basic element to life.
See also section 3.

Question 8

Answer   Prior to 1828, all compounds of carbon were obtained more
or less directly from living or dead organisms. The chemistry
of the element carbon thus became known as organic chemistry.
See also section 3.

Question 9

Answer   See your text, section 4.1.

Question 10

Answer   (i) b; (ii) d.

Comment   It was stated in section 4.5 that one source of chirality in an organic molecule is a tetrahedral carbon atom which is attached to four different groups.

X need not be chiral just because it lacks a plane of symmetry (you can see that it does if you draw a three dimensional representation or use models); it could possess less common forms of symmetry which would mean that it could be superimposed on its mirror image.

The only reliable definition of chiral object is that it cannot be superimposed on its mirror image. A plane of symmetry is the commonest feature which destroys chirality and an object which possesses one is definitely not chiral.

Molecule X is unsaturated. It possesses a multiple bond. Unsaturation has nothing to do with non-bonding electrons which in X are associated with the two oxygen atoms.
See also section 4.

Question 11

Answer   (iii)

Comment   The valencies of all three atoms are satisfied with the formula $O=C=O$. In this structure carbon participates in two identical four electron bonds, the mutual repulsion of which dictates a linear arrangement of the atoms.
See also section 4.

Question 12

Answer   (i)

Comment   Non-bonding pair/double bond interaction is unambiguously the most severe repulsion (section 4.1) so (ii) and (iii) are eliminated. Whether double bond/single bond repulsion is greater than non-bonding pair/single bond repulsion is less clear.
See also section 4.

Question 13

Answer   (i), (v), (vi), a; (ii), (iii), (iv), b.

Comment   The geometry of carbon atoms involved in double bonds is always planar and triple bonds always linear. The carbon atoms in (iv) are not tetrahedral.
See also section 4.

Question 14

Answer   (iii) a ; (iv) b ; (i), (ii), c.

Comment   (A) and (iii) are mirror images, the mirror plane being that of the paper. Furthermore, as (A) possesses a carbon atom bound to four different groups, they are not superimposable.

(iv) can be 'converted' into (A) by a (forbidden) rotation about the $C=C$ bond.

(ii) is not a stereoisomer of (A) because different bonding is involved and (i) incorporates both relationships involved in (iii) and (iv).
See also section 4.

Question 15

Answer   See the key on p. 90.

Question 16

Answer

Question 17

Answer   (I) diamond ; (II) graphite.

Comment   1. Structure II shows an extensive conjugated system which implies high polarizability and this material would thus be expected to absorb visible light very strongly.

In fact, within the carbon atom planes of the graphite structure all C.C. bond distances are equal. Thus, the alternate double/single bond structure is not quite correct but the electrons in such a system would still be highly polarizable and expected to give rise to strong absorption of visible light.

2. I shows strong carbon carbon bonds in every direction whereas II has very long weak bonds perpendicular to the planes containing the carbon atoms. This leads to very easy cleavage or slipping parallel to the planes and accounts for the use of graphite in lubrication and pencil leads.

Structure I by contrast is very rigid and this partly explains the hardness of diamond.

See also section 5.

Key for Self-Assessment Question No. 15

Section of text	(a) 4.4C	(b) 4.2	(c) 4.4A	(d) 4.4C	(e) 4.4A	(f) 4.2
(i) 4.5	/		/		/	
(ii) 4.3	/			/		
(iii) 4.2 and 4.4A		/	/		/	
(iv) 4.4A			/		/	
(v) 4.4C	/			/		
(vi) 4.3			/		/	
(vii) 4.3		/	/		/	/
(viii) 4.5		/		/		/
(ix) 4.2						
(x) 4.5		/		/		/
(xi) 4.4A			/			
(xii) 4.4A			/		/	
(xiii) 4.4C	/			/		
(xiv) 4.5	/		/		/	
(xv) 4.5						

Question 18

Answer   (i) c a b d

Comment   (increasing molecular size of compounds of the same structural type).

Answer   (ii) c b a

Comment   hydrogen bonding and bond polarization lead to large intermolecular forces.

See also section 5.

Question 19

Answer   (i) c b a

Comment   Compounds become less soluble as non-polar hydrocarbon fragments increase in size.

Answer (ii) b c a

Comment Increasing polarity parallels increasing solubility in water. See also section 5.

Question 20

Answer (i) (ii), b ; (iii) (iv), a.

Comment The specificity of enzyme catalysis in this context comes largely from the fact that both it and the molecule it interacts with are chiral. Thus the relationship between them is different from the relationship of the same enzyme to the molecule's mirror image (section 5.5). If the molecule was achiral this difference would disappear. Both a man and a horse (subject to the absence of some deformity) are achiral in external appearance so whether the riding is being done astride or side-saddle, one would not detect any difference if either horse or rider were replaced by their mirror image.

In (iii), however, both a right-hand-drive car and a Continental highway (used within the law) are chiral. There is a very noticeable difference between driving a British car and its mirror image (an export model) on the Continent.

Again in (iv), the stance taken up by both partners in ballroom dancing is chiral and it is doubtful whether either partner would get much enjoyment out of dancing with the other's mirror image.

Question 21

Answer (i), (iv), (ix)

Comment All contain the —OH group.

Answer (ii), (vi)

Comment Both are saturated hydrocarbons.

Answer (iii), (v)

Comment Both contain oxygen linked to two carbon atoms. Chemical reactivity expected to stem from the non-bonding electrons.

Answer (iv), (xi)

Comment Both contain the same type of unsaturated ring.

Answer (vii), (viii), (ix)

Comment All contain one $C=C$ bond.

Answer (ix), (x)

Comment Both contain $C=O$ groups
See also section 5.

Question 22

Answer (i), (ii), b; (iii) a.

Comment See text, sections 4.3, 5.2, 5.4 and 5.5.

Question 23

Answer   (i) b; (ii) f; (iii) e; (iv) a; (v) c; (vi) d.

Comment   (i) contains an extended conjugated system together with non-bonding electrons and is therefore expected to have colour.

              (ii) contains a long non-polar chain and a polar end group.

              (iii) is part of a long chain molecule of polypropylene which you met in the text.

              (iv) is the major constituent of natural gas

              (v) is a sulphonamide

              (vi) is an organochlorine compound.

See also section 6.

Question 24

Answer   (i) synthetic detergents

             (ii) Novocaine

             (iii) variations on structures known to be effective pesticides

             (iv) incorporation of polar groups into highly coloured compounds to effect adhesion to fabrics.

Comment   See section 6.

Answers to In-text Questions

Answer 1, p. 14

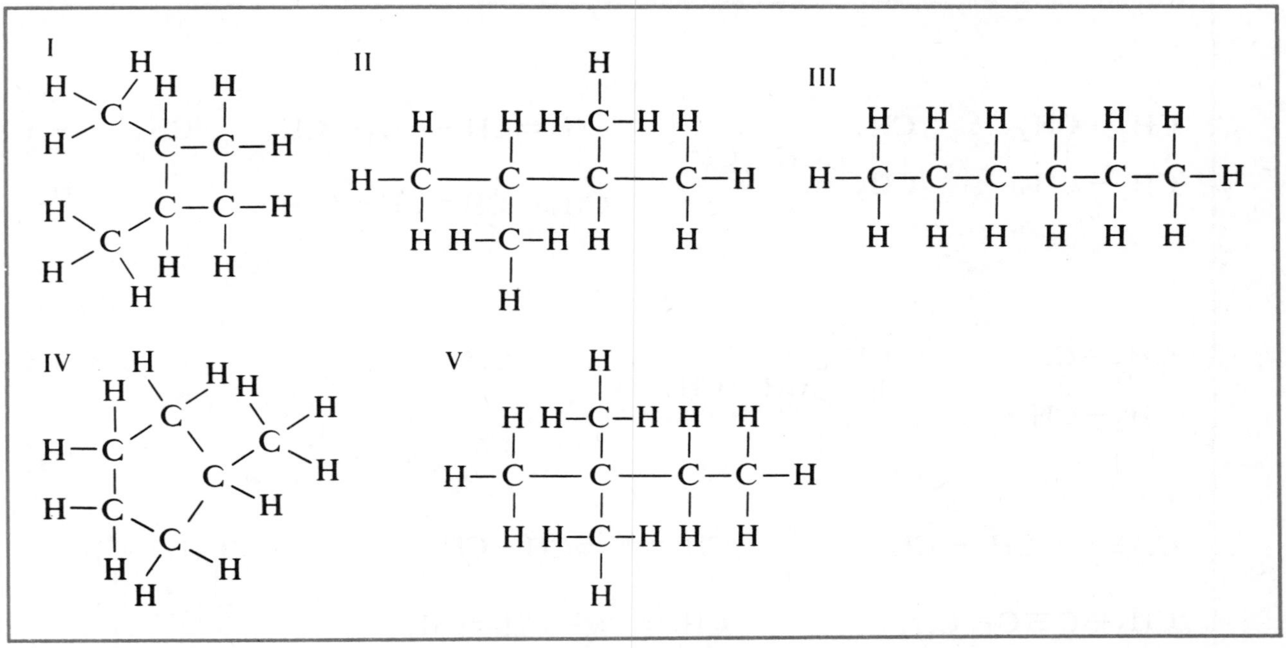

Answer 2, p. 14

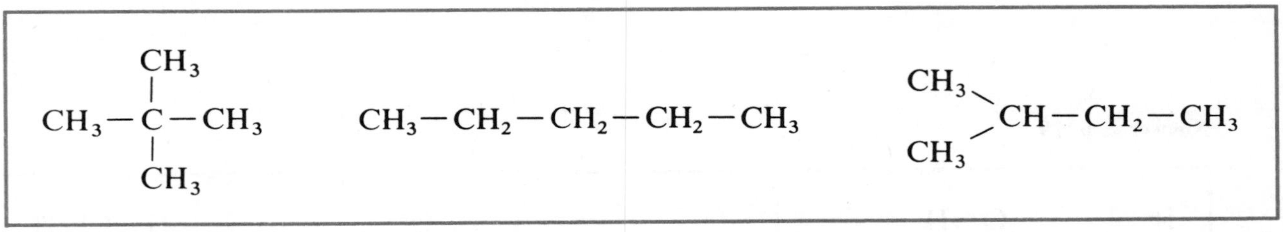

Answer 3, p. 16

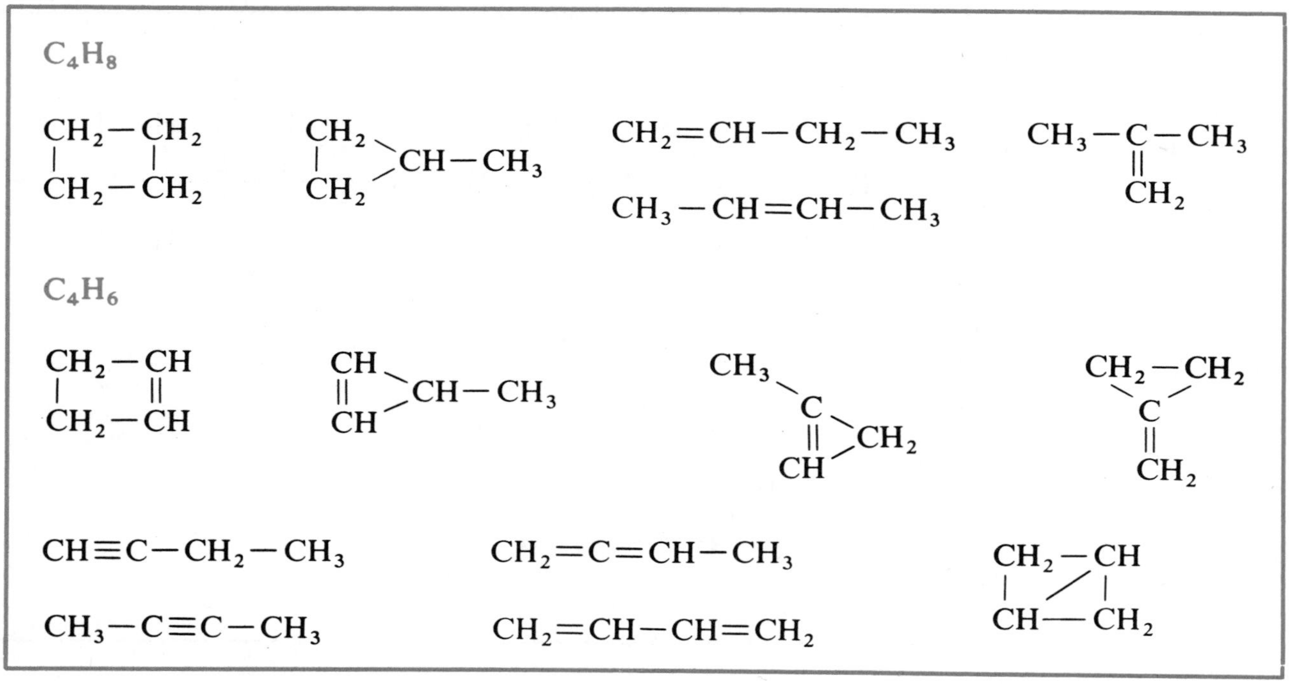

C_4H_8

$$\begin{array}{c} CH_2-CH_2 \\ |\qquad\ | \\ CH_2-CH_2 \end{array}$$

$$\begin{array}{c} CH_2 \\ \big|\quad\diagdown \\ \quad\quad CH-CH_3 \\ \big|\quad\diagup \\ CH_2 \end{array}$$

$$CH_2{=}CH-CH_2-CH_3$$

$$CH_3-CH{=}CH-CH_3$$

$$\begin{array}{c} CH_3-C-CH_3 \\ \| \\ CH_2 \end{array}$$

C_4H_6

$$\begin{array}{c} CH_2-CH \\ \|\qquad\ \| \\ CH_2-CH \end{array}$$

$$\begin{array}{c} CH \\ \|\quad\diagdown \\ \quad\quad CH-CH_3 \\ CH \diagup \end{array}$$

$$\begin{array}{c} CH_3 \\ \diagdown \\ \quad C \\ \diagup\ \| \quad\diagdown CH_2 \\ CH \end{array}$$

$$\begin{array}{c} CH_2-CH_2 \\ \diagdown\quad\diagup \\ C \\ \| \\ CH_2 \end{array}$$

$$CH{\equiv}C-CH_2-CH_3$$

$$CH_3-C{\equiv}C-CH_3$$

$$CH_2{=}C{=}CH-CH_3$$

$$CH_2{=}CH-CH{=}CH_2$$

$$\begin{array}{c} CH_2-CH \\ |\quad\diagup\ | \\ CH-CH_2 \end{array}$$

Answer 4, p. 19

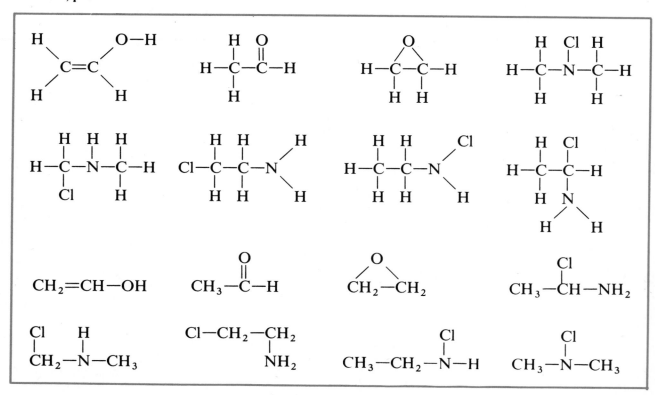

Answer 5, p. 34

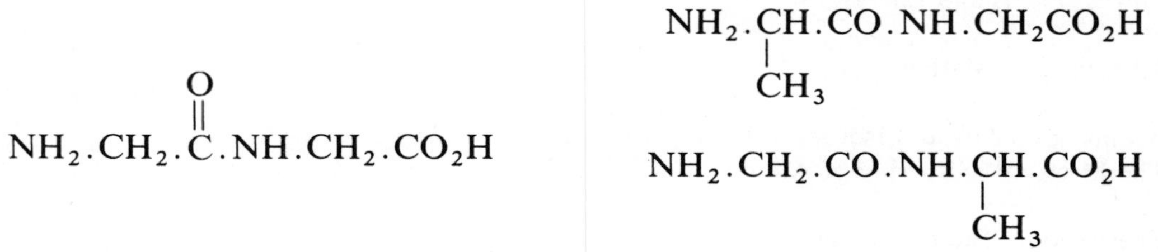

$$NH_2.CH_2.\overset{\displaystyle O}{\overset{\|}{C}}.NH.CH_2.CO_2H$$

$$NH_2.\underset{\underset{\displaystyle CH_3}{|}}{CH}.CO.NH.CH_2CO_2H$$

$$NH_2.CH_2.CO.NH.\underset{\underset{\displaystyle CH_3}{|}}{CH}.CO_2H$$

Answer 6, p. 37

structural
↓
stereo
↙ ↘
geometric   *optical*

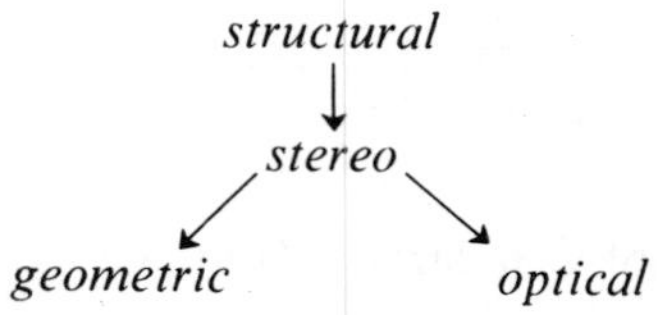

Answer 7, p. 65

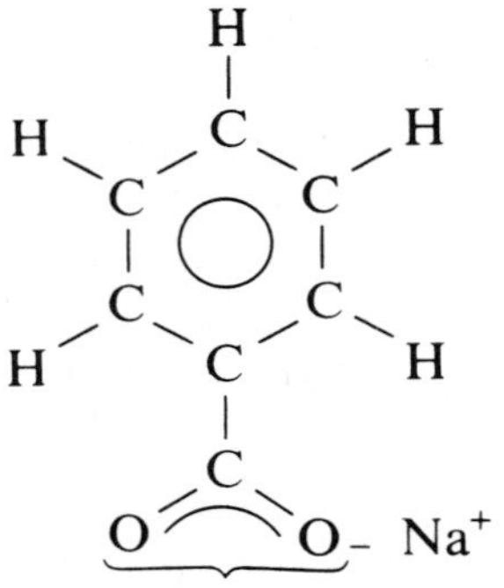

Figure 33   Sodium benzoate.

S.100—SCIENCE FOUNDATION COURSE UNITS

1   Science: Its Origins, Scales and Limitations
2   Observation and Measurement

3   Mass, Length and Time
4   Forces, Fields and Energy

5   The States of Matter

6   Atoms, Elements and Isotopes: Atomic Structure
7   The Electronic Structure of Atoms

8   The Periodic Table and Chemical Bonding
9   Ions in Solution

10   Covalent Compounds

11 ⎫
12 ⎭ Chemical Reactions

13   Giant Molecules

14   The Chemistry and Structure of the Cell

15 ⎫
16 ⎭ Cell Dynamics and the Control of Cellular Activity

17   The Genetic Code: Growth and Replication
18   Cells and Organisms

19   Evolution by Natural Selection
20   Species and Populations

21   Unity and Diversity

22   The Earth: Its Shape, Internal Structure and Composition

23   The Earth's Magnetic Field

24   Major Features of the Earth's Surface
25   Continental Movement, Sea-floor Spreading and Plate Tectonics

26 ⎫
27 ⎭ Earth History

28   The Wave Nature of Light

29   Quantum Theory
30   Quantum Physics and the Atom

31   The Nucleus of the Atom
32   Elementary Particles

33 ⎫
34 ⎭ Science and Society

The most stable.

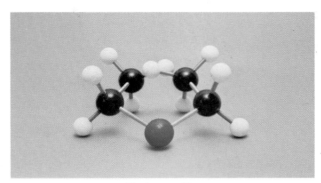

The least stable.

Figure 8   Two conformations of diethyl ether.

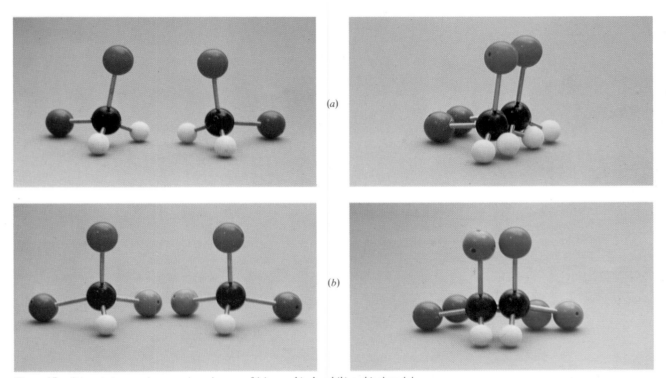

(a)

(b)

Figure 17   Attempts to superimpose mirror images of (a) an achiral and (b) a chiral model.

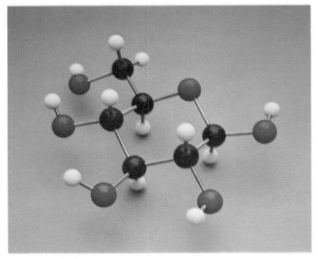

Figure 14   (+) Glucose.

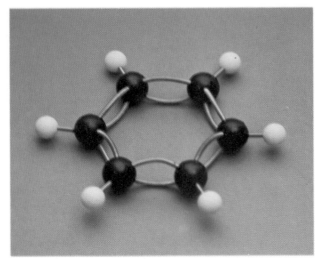

Figure 32   Benzene.